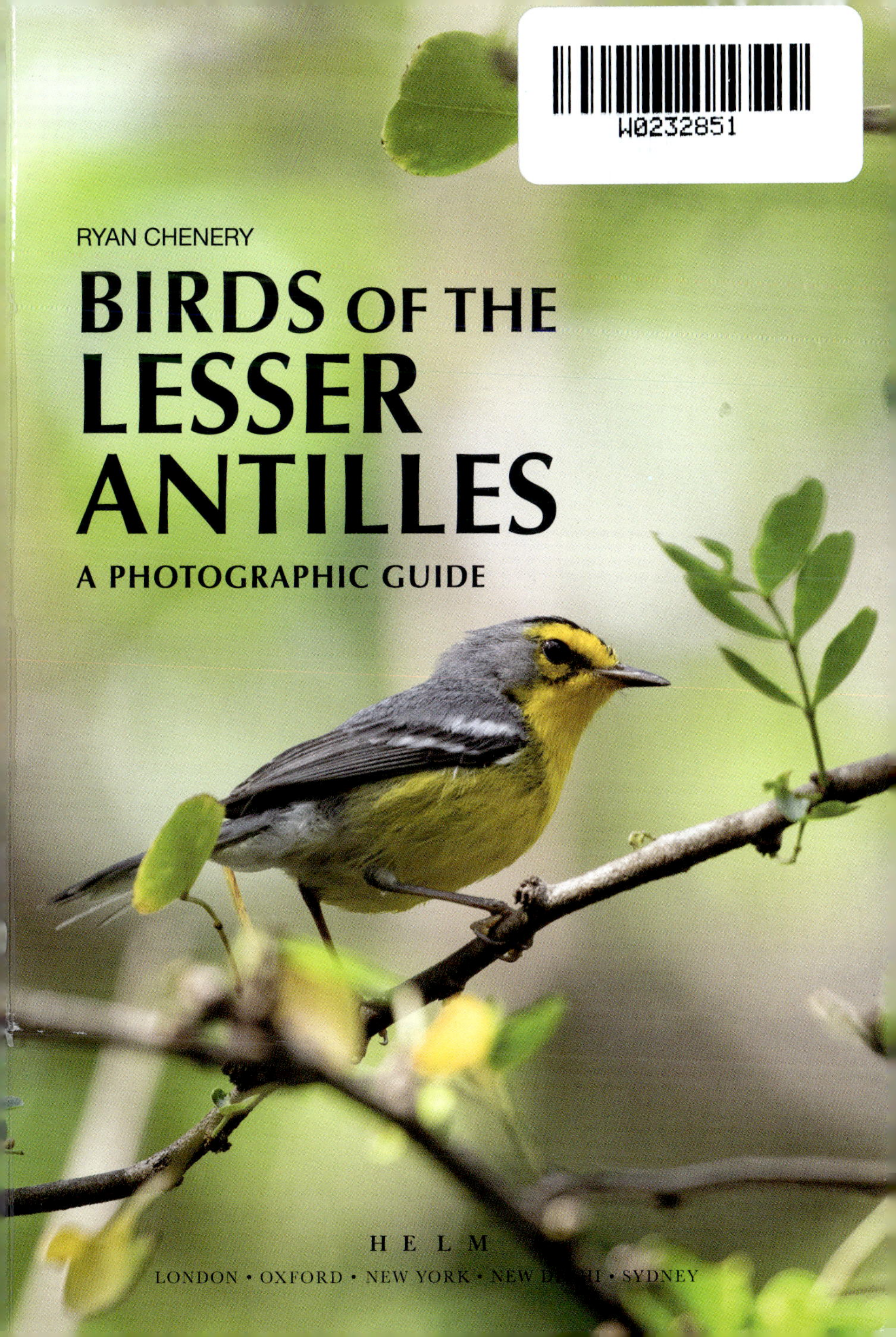

RYAN CHENERY

BIRDS OF THE LESSER ANTILLES

A PHOTOGRAPHIC GUIDE

HELM
LONDON • OXFORD • NEW YORK • NEW DELHI • SYDNEY

To my wonderful parents Percy and Jeanne for their unwavering support, and for encouraging me to always follow my dreams; and to my Lil Family (my beautiful wife Alex and two precious daughters Ariadne and Leilani): birding may be my passion, but the three of you are unquestionably my heart. Thank you all for your constant love and belief in me, and for allowing holidays to always be in places where we can fit in a bit of birding on the side!

HELM
Bloomsbury Publishing Plc
50 Bedford Square, London, WC1B 3DP, UK
29 Earlsfort Terrace, Dublin 2, Ireland

BLOOMSBURY, HELM and the Diana logo are trademarks of Bloomsbury Publishing Plc

First published in the United Kingdom 2022

A catalogue record for this book is available from the British Library.
Library of Congress Cataloguing-in-Publication data has been applied for.

ISBN: PB: 978-1-4729-8961-1; ePub: 978-1-4729-8960-4;
ePDF: 978-1-3994-0516-4

2 4 6 8 10 9 7 5 3 1

Design by Susan McIntyre
Printed and bound in China by RR Donnelley Asia Printing Solutions Limited Company

MIX
Paper from responsible sources
FSC® C144853
FSC www.fsc.org

CONTENTS

Acknowledgements	4
Introduction	5
Birdlife on the Lesser Antilles	6
Bird Conservation	6
Maps of the Region	8
Key Birdwatching Habitats	9
Key Birdwatching Sites on Each Island	13
Species Accounts	24
Glossary of Terms	220
Bibliography	221
Photo Credits	222
Index	223

Acknowledgements

A special and most heartfelt thanks to Patricia E. Bradley (the author of *Birds of the Cayman Islands* and co-author of *Birds of Cuba: A Photographic Guide*), and the late Yves-Jacques Rey-Millet (photographer for *Birds of the Cayman Islands* and *Birds of Cuba: A Photographic Guide*) for granting permission to use a selection of Yves-Jacques' photos, and for their encouragement of my birding tour company (Birding the Islands: www.birdingtheislands.com) that now delivers birdwatching trips throughout the Lesser Antilles.

I would also like to extend my deepest thanks to all of the other photographers who kindly allowed me to include their fabulous images in this book. A personal thank you to: Alex Large, Alexandra Chenery, Alicia Williams, Andrea Easter-Pilcher, Anthony Levesque, Béatrice Henricot, David Petts, Faraaz Abdool, George Tuthill, Jane Hartline, John Dyson, Julian Moore, Keith Clarkson, Larry Therrien, Mark Greenfield, Mike Barth, Nigel Lallsingh, Paul R. Reillo, Skye Haas, Steve Race, Steven Woon, Vaughan Ashby and Vaughan Francis. A special thanks also to the Saint Vincent and the Grenadines Tourism Authority for allowing me to use some of their wonderful images.

Thank you to Bertrand Jr. Baptiste, Lester Nanan and Lisa Sorenson for their recommendations of photographers to contact.

Thank you to my publisher, Jim Martin, for approaching me to undertake this work. It has been a project I have enjoyed immensely, and I am grateful to him for the opportunity. Thank you also to my editor, David Campbell, for his advice and assistance with format, layout and content.

White-headed Munia.

INTRODUCTION

The islands of the Lesser Antilles are renowned worldwide for picturesque beaches, stunning coastlines and sparkling turquoise seas – and they do indeed offer such attractions to the hundreds of thousands of visitors who visit the region every year. But look closer, and you will find that they are also home to a host of spectacular birds, many of which are threatened and often found nowhere else on the planet.

Whether you are heading off on vacation to relax on the beautiful white sand beaches of Antigua and Barbuda; on a hiking holiday through the verdant forests of the 'Nature Lovers' Island' of Dominica; surfing the thunderous rollers that batter the rugged east coast of Barbados; or if you are one of the increasing number of birders targeting these islands as an adventurous and rewarding destination to add to your life list, this book will help you identify the many feathered gems to be found along this remarkable island chain.

Nestled within the easternmost part of the Caribbean, the Lesser Antilles comprises a chain of islands which, running from north to south, begins in Anguilla (18°13′23.77″ N 63°03′23.88″ W) and descends in a long sweeping arc culminating on the 'Spice Isle' of Grenada (12°06′35.89″ N 61°41′36.66″ W). Ideally positioned between North and South America, with the warm tranquil waters of the Caribbean Sea bordering their western coastlines and the heaving Atlantic Ocean thundering against their eastern shores, these multiple island nations are ideally situated to receive flights of birds: almost unfathomable numbers of North American migrants during autumn and spring; Old World vagrants that have inadvertently crossed the Atlantic; and even austral migrants from South America that have set off on an exploratory journey north (much like the original human colonisers of the islands did some 5,000 years ago). Added to these fascinating visitors is a wondrous array of regional breeding residents, including a wealth of endemics and near-endemics, and a host of endemic subspecies, all of which serves to ensure your birding experience in the region will be truly memorable.

BIRDLIFE ON THE LESSER ANTILLES

Species present year-round

Endemics, e.g. Imperial Parrot on Dominica, Whistling Warbler on St Vincent, Montserrat Oriole on Montserrat.

Near-endemics, e.g. Blue-headed Hummingbird on Martinique and Dominica, Plumbeous Warbler on Guadeloupe and Dominica, Grenada Flycatcher on Grenada and St Vincent, Grey Trembler on St Lucia and Martinique.

Lesser Antillean Endemics – species restricted to the Lesser Antillean region, where they are found on multiple islands, e.g. Purple-throated Carib, Scaly-breasted Thrasher, Brown Trembler, Lesser Antillean Saltator.

Breeding Residents – species that do not migrate outside of the region post-breeding, e.g. Caribbean Elaenia, Scaly-naped Pigeon, Common Ground Dove.

Species primarily present in the Lesser Antilles during spring migration (March and April) and autumn migration (July to October)

Migrant waders, waterfowl, herons and warblers, along with other migratory species such as kingfishers, grosbeaks, cuckoos and buntings.

Species primarily present in the Lesser Antilles during the summer months (May to September)

The number of terns and gulls increases significantly during summer, with several species breeding on rocky islets in the region and others merely visiting.

Migratory species present in the Lesser Antilles during the North American winter (November to February)

Although typically not in very large numbers, individuals from a variety of migratory species overwinter in the region. Therefore, a selection of migrant waders, warblers and herons, among others, can be seen on the islands between November and February.

Bird conservation

Although very popular as vacation destinations, the islands of the Lesser Antilles are not as well known in birding circles, and are often overshadowed by their larger Caribbean cousins – the Greater Antilles. These relatively gargantuan neighbours have also historically received more attention from researchers and conservation bodies. However, in recent years the focus has begun to shift, with increased research revealing that urgent action is needed on these tiny islands to prevent several species from being lost forever.

When the global population of a species is restricted to a single tiny island of 134km^2, it is already fighting an uphill battle for survival. But when compounded by the said island's location within a hurricane belt, destruction of crucial habitat through construction projects and land development, and predation by an invasive species, the odds of survival plummet significantly. Such is the predicament facing the demure and retiring Grenada Dove; now clinging to survival with a fragmented cumulative population of some 136 individuals remaining in two tiny pockets of dry coastal forest on Grenada. Sadly, the plight of this species does not represent an isolated case study in the Lesser Antilles – indeed, shockingly, it may not even be the most Critically Endangered endemic bird in the region. Such a lamentable distinction could well be held by the majestic Imperial Parrot on Dominica, where deep in the dense and cloud-enshrouded montane forests of the 'Nature Isle' an unknown number continue to battle against the threats posed by foes both natural (hurricanes like Maria in 2017) and man-made (the illegal pet trade). Throughout the region, several other endemics such as the St Vincent Parrot, Whistling Warbler, St Lucia Black Finch, Barbuda Warbler and Montserrat Oriole also walk a precarious line between existence and oblivion.

Yet, there is reason for hope and cause for optimism. Indeed, for inspiration, one need look no further than the success achieved on St Lucia, with her own endemic parrot's return from the brink. In the 1970s, following years of habitat destruction, hunting and capture for the illegal pet trade, the global population of the St Lucia Parrot fell to a paltry 100 individuals. However, following an intensive outreach and public awareness program by the St Lucia Department of Forestry, the establishment of a parrot reserve, and a ban on hunting, numbers have risen dramatically to the point where these beautiful parrots can once again be regularly seen coursing across the skies – every mesmeric beat of their spectacularly colourful wings against the unbroken backdrop of verdant rainforest, a resounding and emphatic reminder of what can be achieved when we make a concerted effort to preserve rather than destroy. Such success also underlies the critical importance of the continued efforts of organisations like the Rare Species Conservatory Foundation (providing crucial support to wildlife conservation on Dominica); conservation charities such as BirdsCaribbean, who raise awareness and empower local partners to build a region where people appreciate, conserve and benefit from thriving bird populations and ecosystems; and indeed to birding tour operators such as Birding the Islands, who directly support conservation efforts in the region through donations from their tours. The ability of organisations to speak and act on behalf of those who cannot, often means the difference between life and death for threatened species. For some birds in the Lesser Antilles, the clock is well and truly ticking.

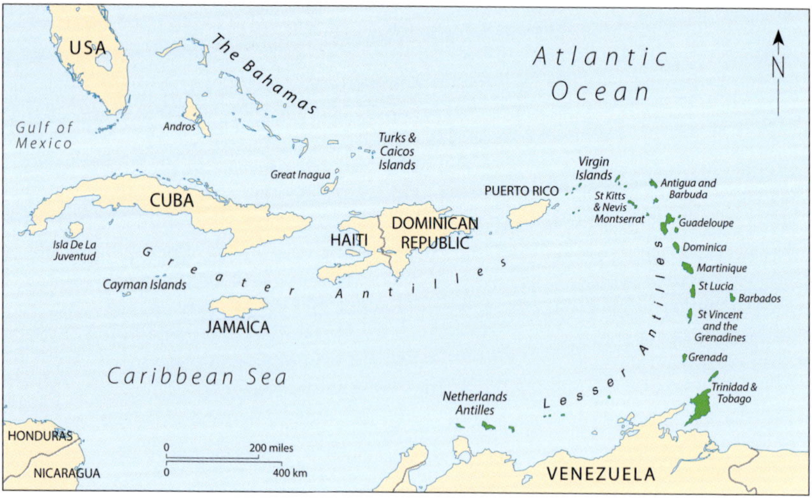
USA
The Bahamas
Atlantic Ocean
N
Gulf of Mexico
Andros
Turks & Caicos Islands
Great Inagua
Virgin Islands
PUERTO RICO
Antigua and Barbuda
St Kitts & Nevis
Montserrat
CUBA
DOMINICAN REPUBLIC
HAITI
Guadeloupe
Dominica
Isla De La Juventud
Greater Antilles
Martinique
St Lucia
Barbados
Cayman Islands
St Vincent and the Grenadines
JAMAICA
Lesser Antilles
Grenada
Caribbean Sea
Trinidad & Tobago
Netherlands Antilles
HONDURAS
0
200 miles
0
400 km
NICARAGUA
VENEZUELA

British Virgin Islands
Puerto Rico
Anguilla
Virgin Islands
N
Antigua and Barbuda
St Kitts and Nevis
Montserrat
Atlantic Ocean
Guadeloupe
Lesser Antilles
Dominica
Caribbean Sea
Martinique
St Lucia
Barbados
St Vincent and the Grenadines
Grenada
Trinidad and Tobago
0
150 miles
0
300 km
Venezuela

KEY BIRDWATCHING HABITATS

Coast

Be sure to scan the ocean for the terns, gulls, boobies, pelicans and frigatebirds that are drawn primarily to southern and western shorelines. Check northern and eastern shores for shearwaters, skuas (called jaegers in North America), petrels and tropicbirds. Also explore beaches, tidal pools and exposed reefs for foraging waders and herons. Remote beaches also represent ideal nesting sites for resident waders such as Wilson's Plover.

Dry scrubland

On first glance, dry scrubland may not look particularly appealing as a wildlife habitat, but being largely composed of grasses, shrubs and drought-tolerant trees (most with a compound leaf structure), it can often be very rewarding for a number of flycatcher, wren, mockingbird, cuckoo, warbler, grassquit, nightjar and dove species.

Woodland (dry broadleaved forest)

A woodland patch filled with deciduous tree species is a haven for warblers in temperate regions, and the same is true in the tropics. A variety of migrant warblers seek out such sites while on the islands. Other species that thrive in this habitat include orioles, vireos, pigeons, doves and saltators. Be sure to scan the understorey carefully for wrens, thrashers and finches.

Wetlands

One of the most productive habitats for birding in the region. Mangrove swamps, saltmarshes, lakes and lily ponds all attract a plethora of species, including migrant waders and waterfowl, kingfishers and herons, and other aquatic birds such as grebes, gallinules and rails.

Montane forest

This is typically one of the more challenging habitats to access, but unquestionably one of the most rewarding in terms of possible sightings. With species of parrot, hummingbird, solitaire, euphonia, thrush, oriole and quail-dove often restricted to these higher elevations, it is well worth the effort of planning a visit.

Habitat edges

The point at which one habitat ends and another habitat begins can produce numerous species. For example, scan the treeline where woodland ends and grassland begins for thrashers, tremblers, warblers, flycatchers, hummingbirds and falcons.

Grassland

Vast expanses of native grassland are few and far between on most islands. However, the patches that remain are well worth visiting. Numerous species of swallow, finch, flycatcher and dove can thrive there.

Tropical forest

Ranging from lush, moist belts, to low-lying, sweltering, humid expanses, this most biodiverse of habitats in the region is home to a vast array of species. These include drab musicians such as Cocoa Thrush and House Wren but also showy characters like Montserrat Oriole and St Lucia Parrot.

KEY BIRDWATCHING SITES ON EACH ISLAND

ANGUILLA

EAST

East End Pond
(18°14′11.4″N 62°59′49.8″W)

Designated an Important Bird Area (IBA) by BirdLife International, this 5-ha saltwater pond with its expanse of Button and Red Mangroves is a mecca for large flocks of migrant waders. These can include Willet, Greater and Lesser Yellowlegs, Short-billed Dowitcher, Stilt Sandpiper, Black-necked Stilt, Snowy Plover, Semipalmated Plover and Kildeer. Resident and migrant waterfowl frequent the site, including Lesser Scaup, Ruddy Duck, White-cheeked Pintail, Blue-winged Teal, Ring-necked Duck, Northern Shoveler and American Wigeon. A variety of heron species nest there, while others use it as a feeding and rest stop during migration. These can include Great Egret, Green Heron, Tricolored Heron and Yellow-crowned and Black-crowned Night-herons.

NORTH

Prickly Pear East
(18°15′52.4″N 63°10′18.2″W)
and **Prickly Pear West**
(18°16′18.6″N 63°11′14.4″W)

The cays located just 8 km north of Anguilla represent a haven for seabirds. A visit there will almost certainly provide encounters with such species as Brown Booby, Red-billed Tropicbird, Magnificent Frigatebird, Sooty Tern, Bridled Tern, Least Tern, Laughing Gull, Brown Pelican and an array of other ocean wanderers.

SAINT MARTIN/SINT MAARTEN

Saint Martin (the French side)

CENTRAL

Pic Paradis (18°04′30.2″N 63°02′58.0″W)

This is the highest peak on the island and supports the dry forest habitat that is home to numerous regional specialities such as Pearly-eyed Thrasher, Zenaida Dove, Grey Kingbird, Purple-throated Carib, Antillean Crested Hummingbird, Black-whiskered Vireo and Lesser Antillean Bullfinch. Also note the attractiveness of such habitat to migrant warblers, including Northern Parula, Black-and-white Warbler, Cape May Warbler and American Redstart.

EAST

Salines d'Orient
(18°04′50.0″N 63°01′06.5″W)

Lying to the west of Orient Bay is a vast wetland area offering one of the best sites on the island for encountering a veritable barrage of waders, including White-rumped Sandpiper, Snowy Plover, Grey Plover, Least Sandpiper, Spotted Sandpiper, American Golden Plover, Sanderling, Ruddy Turnstone and Short-billed Dowitcher. Its proximity to the bay allows superb opportunities to observe Brown and Red-footed Boobies, along with several tern species, including Least, Sandwich and Royal.

Sint Maarten (the Dutch side)

CENTRAL

Great Salt Pond
(18°01′58.6″N 63°03′01.4″W)

A large lake and wetland area that attracts a host of migrant herons, Great Salt Pond is also of interest for numerous species of gull and tern due to its size. These include Roseate, Least and Royal Terns, along with Laughing Gull and the seldom seen Lesser Black-backed and Ring-billed Gulls. Both Pied-billed Grebe and Black-necked Stilt nest there, and the habitat is ideal for many migrant waders, including Semipalmated, Least and Pectoral Sandpipers and Whimbrel. A wide selection of regional species can also be found in habitat surrounding the pond – including Green-throated Carib and Pearly-eyed Thrasher – and the site is highly attractive to migrant warblers such as Magnolia and Palm Warblers and Northern Parula. The open water also draws in significant numbers of Barn, Cliff and Cave Swallows, along with Caribbean Martin.

ANTIGUA AND BARBUDA

Island of Antigua

NORTH

McKinnon's Salt Pond
(17°08′53.9″N 61°51′05.2″W)

Without question the best location on the island for views of herons, waders and waterfowl that are all reliant on this large mangrove waterbody located on the north-west coast. With such an optimum location on one of the more northerly of the Lesser Antillean islands, the selection of migrant waterfowl species can be wider here than on many of the islands further south, and includes Northern Pintail, Northern Shoveler and Ring-necked Duck. There is also the strong possibility (especially at dawn and dusk) for sightings of the rare breeding resident West Indian Whistling Duck and the more regionally common White-cheeked Pintail. Both Black-necked Stilt and Wilson's Plover nest on the sandy banks, and several migrant wader species forage in the shallow ponds. McKinnon's is a good site for migrant warblers with an affinity for habitat near water, and regular visitors include Northern Waterthrush, Blackpoll Warbler, American Redstart and Northern Parula. 'Golden Warbler', the resident subspecies of Yellow Warbler, is also present. Other year-round species of note in the area are White-crowned Pigeon and White-winged Dove, along with Common Gallinule, American Coot, Lesser Antillean Flycatcher, Pied-billed Grebe, American Kestrel and Black-faced Grassquit.

Island of Barbuda

WEST

River Road (17°36′11.1″N 61°49′44.7″W)

Barbuda is a very small and largely undeveloped island with one main road, Route 1 (also known as River Road), running from north to south. If you drive in either direction along this road and pause at any pond or body of water near the roadside, you have a good chance at the appropriate time of year of seeing a few migrant waders. These can include Solitary and Stilt Sandpipers, Short-billed Dowitcher and several other species. Waterfowl in the larger ponds can include White-cheeked Pintail, Blue-winged Teal and American Wigeon. The scrubby vegetation and sparse woodland surrounding the waterbodies (especially in the vicinity of the Sir McChesney George Secondary School) should reveal a host of resident species, including the fairly common endemic Barbuda Warbler, as well as Black-faced Grassquit, Lesser Antillean Flycatcher, Mangrove Cuckoo, Common Ground Dove, Caribbean Elaenia, White-winged Dove and Black-whiskered Vireo. Migrant warblers can also be found in such habitat.

NORTH

Codrington Lagoon National Park
(17°41′41.6″N 61°52′07.4″W)

Take a short boat ride from the city of Codrington to enjoy the truly amazing spectacle of being up close and personal with the colony of thousands of Magnificent Frigatebirds that nest in the mangroves bordering the lagoon. Also possible there are resident and migrant heron species such as Great Egret, Little Blue Heron, Black-crowned and Yellow-crowned Night-herons, and a selection of terns and gulls.

ST KITTS AND NEVIS

Island of St Kitts

SOUTH

Frigate Bay Pond
(17°17′02.2″N 62°41′17.4″W)

This stunning lagoon can be a very productive site for waders such as Greater and Lesser Yellowlegs and Stilt Sandpiper, along with a host of smaller species including Spotted, Semipalmated and Western Sandpipers. Migrant and resident herons include Great Blue, Little Blue and Green Herons and Great Egret. Also drawn to this vast expanse of water are Common Gallinule, American Coot, Osprey, Belted Kingfisher and significant numbers of Black-necked Stilts. Other less numerous aquatic species can include Sora, Glossy Ibis and the very rarely seen Clapper Rail. The surrounding landscape is attractive for several non-aquatic breeding

The Green-throated Carib is a large species of hummingbird that is common throughout the region. Seen here feeding at a Heliconia *flower.*

residents, including Pearly-eyed Thrasher, Lesser Antillean Bullfinch, Black-faced Grassquit, Scaly-breasted Munia and Caribbean Elaenia.

Island of Nevis

CENTRAL

Golden Rock rainforest
(17°08′35.4″N 62°34′00.8″W)

To the south-east of towering Nevis Peak in St George Gingerland Parish is a lush forest. This is a site that represents a golden opportunity for birding. Drawn to such habitat are a host of species, including Scaly-naped Pigeon, Purple-throated Carib, Green-throated Carib, Bridled Quail-dove, Scaly-breasted Thrasher, Lesser Antillean Flycatcher and even the majestic Red-tailed Hawk. This site is also sought out by migrant warblers such as American Redstart and Black-and-white and Hooded Warblers.

MONTSERRAT

NORTH

Brades (16°47′34″N 62°12′38″W) and **St Peter Parish**

Birding on Montserrat is restricted to the north of the island, as vast areas of the south are located within the exclusion zone, an area deemed unsafe due to the threat of volcanic activity. Thankfully, Montserrat is one of the most densely forested islands in the region and a system of trails (some of which are signposted) run throughout these forests. Many of these (such as Blackwood Allen Hiking Trail, Dry Waterfall Trail and Oriole Trail) run through St Peter Parish in the lush mountainous areas south of Barzeys and Cudjoehead. Birding on the trails can be very rewarding and you will have opportunities for sightings of the endemic Montserrat Oriole and other specialities such as Bridled Quail-dove and Forest Thrush, as well as more commonly seen forest inhabitants, including Antillean Crested Hummingbird, Scaly-breasted Thrasher, Lesser Antillean Bullfinch, Scaly-naped Pigeon, Purple-throated Carib and Pearly-eyed Thrasher. These forests are also the favoured haunts of many migrant species of North American warbler, including Louisiana Waterthrush, Magnolia and Black-throated Green Warblers and several others.

Little Bay (16°48′10.4″N 62°12′17.6″W)

This coastal port town with a jetty and sheltered bay is a good site for encounters with Red-billed Tropicbird, Brown Pelican, Magnificent Frigatebird, Brown Booby and several species of gull and tern.

GUADELOUPE

EAST

Pointe des Chateaux and La Grande Saline salt pond (16°14′48.0″N 61°10′39.0″W)

A great area to set up your scope and survey the ocean for seabirds is Pointe des Chateaux. This spit of land extends into the Atlantic and is located near the town of St François. Possibilities here include several tern species (Roseate, Common, Least, Bridled and others), Laughing and Lesser Black-backed Gulls, three species of jaeger (Pomarine, Long-tailed and Parasitic) and three species of booby (Brown, Masked and Red-footed). Keep an eye out, too, for Brown Noddy, Cory's, Audubon's and Manx Shearwaters, South Polar Skua, Red-billed and White-tailed Tropicbirds, Magnificent Frigatebird and Brown Pelican.

La Grande Saline salt pond is a fabulous location for observing migrant waders. Species can include Wilson's Plover and Least, Semipalmated and Solitary Sandpipers. Several introduced species that now breed on the island can also be found there, such as Scaly-breasted Munia and Orange-cheeked and Black-rumped Waxbills. During autumn migration, the vegetated and wooded areas attract migrant warblers including Blackpoll and Northern Waterthrush, along with other visitors such as Yellow-billed Cuckoo. Cliff, Cave and Barn Swallows are also known to mass at the site.

CENTRAL

Parc National de la Guadeloupe

(16°10′47.0″N 61°40′50.0″W)

Located just off of Highway D23 in Parc National de la Guadeloupe, and about a 30-minute drive west from Petit Bourg, is the Cascade aux Ecrevisses. This spectacular setting in the heart of dense primary forest, with its flowing streams and waterfalls, even has picnic tables for you to relax and bird while seated. If you arrive at this location at, or just before, dawn, you will be well placed to enjoy encounters with two of the most elusive species in the Lesser Antillean region: Forest Thrush and Bridled Quail-dove. Other noteworthy species regularly seen at this site include the endemic Guadeloupe Woodpecker, near-endemic Plumbeous Warbler, Pearly-eyed Thrasher, Brown Trembler, Lesser Antillean Pewee, Lesser Antillean Flycatcher and Lesser Antillean Swift (keep a close watch on the skies overhead). Several migrant warblers can also be seen in the forests of Parc National de la Guadeloupe, including Hooded and Chestnut-sided Warblers and American Redstart.

A Guadeloupe Woodpecker emerging from a nest cavity in a palm tree.

NORTH

Barrage de Gaschet

(16°25′10.9″N 61°28′52.7″W)

For migrant waders, waterfowl and herons, be sure to check out Barrage de Gaschet, located to the west of Highway N8. During the North American autumn and winter, the expansive waterbodies there are favoured by migrant Greater and Lesser Yellowlegs, Short-billed Dowitcher, Stilt and Pectoral Sandpipers, Semipalmated and Grey Plovers and other waders. Species of duck that can be seen include Blue-winged Teal, American Wigeon, Ruddy Duck and Black-bellied Whistling Duck. Numerous heron species also frequent the site, including Little Blue, Great Blue, Tricolored and Green Herons, Great Egret and Least Bittern. Other aquatic birds such as American Coot, Sora and Common and Purple Gallinules are also possible. Raptors can include Osprey and Peregrine Falcon. Other species to be seen are the introduced Red Avadavat and Scaly-breasted Munia, along with migrant warblers such as Blackpoll and Yellow Warblers.

DOMINICA

NORTH

Morne Diablotin National Park (15°30′34.9″N 61°25′18.0″W)

The Syndicate Nature Trail that runs through this vast National Park, with its towering peaks and rivers flowing through large tracts of tropical forest and montane forest, is the best site on the island to encounter the critically endangered Imperial Parrot. The largest population of this endemic can be found in the dense forests on the north-eastern slopes of the park. Also present within the park and often seen along the Syndicate Trail is the other species of endemic *Amazona*, the stunning Red-necked Parrot. The trail also provides excellent opportunities for sightings of other breeding residents, including the near-endemic Blue-headed Hummingbird, Plumbeous Warbler and Brown Trembler, the endemic subspecies of House Wren, Pearly-eyed Thrasher, Scaly-breasted Thrasher, Lesser Antillean Pewee, Lesser Antillean Saltator, Broad-winged Hawk and (especially at dawn and dusk) Forest and Red-legged Thrushes. If you are exceptionally fortunate, in the late evening you may even see Black-capped Petrel winging their way back to nesting sites that are suspected at high elevations in this park. Note that this expansive forested region also attracts migrant passerines (especially during autumn), including Scarlet Tanager and Rose-breasted Grosbeak.

WEST

Layou River (15°23′36.0″N 61°25′29.9″W)

During the North American autumn and winter months, the mouth and banks of this river can be especially productive for viewing a wide range of

The typically shy, forest-dwelling Bridled Quail-dove, is found on the islands from St Lucia north along the chain.

migrant waders, such as Greater and Lesser Yellowlegs, Willet, Spotted and Semipalmated Sandpipers, and Grey and Semipalmated Plovers. Kingfishers (both the resident Ringed and migrant Belted) and herons also patrol the banks of the river. The surrounding landscape provides good opportunities for viewing several regional specialities, such as Black-faced Grassquit, Antillean Crested Hummingbird, Carib Grackle and Lesser Antillean Bullfinch. The coastal area is also excellent for sightings of Brown Pelican, Royal Tern, Laughing Gull and several other seabirds.

MARTINIQUE

SOUTH

L'Etang des Salines (14°24′02.2″N 60°52′24.2″W)

Near the town of St Anne, take some time to visit the saltwater marsh of Etang des Salines for opportunities to see a host of migrant waders, including Semipalmated, Western and Spotted Sandpipers, Whimbrel, Wilson's Plover and Willet. In the surrounding landscape, there is also a good chance to see several of the introduced species that now breed on the island, including Orange-cheeked, Black-rumped and Common Waxbills. Other breeding residents found regularly in this area include Carib Grackle, Zenaida and Common Ground Doves, Tropical Mockingbird, Spectacled Thrush, 'Golden Warbler', Grassland Yellow-finch and Eared Dove.

NORTH AND CENTRAL

Forêt Pitons du Carbet (14°41′39.7″N 61°05′11.9″W)

Located south of Fonds-Saint-Denis, off Highway N3, the Carbet Mountains are the most ancient on the island, and the lush, forested habitat, with its deep gullies and rivers, is one of the prime birding sites on the island. It provides ideal habitat for forest-loving birds such as the endemic Martinique Oriole, near-endemic Blue-headed Hummingbird, Purple-throated Carib, Broad-winged Hawk, Lesser Antillean Flycatcher, Rufous-throated Solitaire, near-endemic Grey Trembler, Lesser Antillean Pewee, Spectacled Thrush, Lesser Antillean Euphonia and Ruddy Quail-dove.

EAST

Reserve Naturelle de la Caravelle (14°46′20.3″N 60°52′59.6″W)

The primarily dry forests and scrub of Presquiîle de la Caravelle represent some of the best sites on the island to see the near-endemic White-breasted Thrasher. Other specialities possible here include Lesser Antillean Saltator, Mangrove Cuckoo, Tropical Mockingbird, White-tailed Nightjar, Ruddy Quail-dove, Green-throated Carib, Black-whiskered Vireo, Caribbean Elaenia and Lesser Antillean Bullfinch. Seabirds along the coast include Magnificent Frigatebird, Royal, Roseate and Sandwich Terns, Brown Noddy and others. In autumn, migrant passerines such as Northern Waterthrush, Blackpoll, Kentucky and Prothonotary Warblers and Yellow-throated Vireo are possible.

ST LUCIA

SOUTH AND EAST

Mon Repos (13°52′48.1″N 60°53′49.5″W)

The dry broadleaved forests lining the coast road from Mon Repos to Dennery Bay harbour an array of species especially fond of this dry, scrubby habitat, and include the endemic St Lucia Black Finch, St Lucia Oriole and St Lucia Pewee. The rare, near-endemic, White-breasted Thrasher, as well as Lesser Antillean Saltator, Mangrove Cuckoo, Bananaquit, Tropical Mockingbird, Antillean Crested Hummingbird, Grey Kingbird and Zenaida and Common Ground Doves can also all be found here. Further north along the east coast, open grassland areas can be rewarding sites for Rufous (aka St Lucia) Nightjar.

Moule-à-Chique (13°42′40.6″N 60°56′31.1″W)

From the plateau overlooking the international airport, watch for Red-billed Tropicbirds whistling by below you, as they visit their nesting sites along the cliff face. American Kestrel and Lesser Antillean Swift are sometimes sighted in the skies over the lighthouse. The wooded area at the base of the hill is also a good site for Scaly-breasted Thrasher and the endemic St Lucia Warbler.

WEST

Soufrière (13°51′24.6″N 61°03′32.7″W)

The blissful historic town of Soufrière, tucked away beneath the towering twin peaks of Les Pitons, is a good site at which to sit and watch the skies above for Magnificent Frigatebirds, Laughing Gulls, and Royal Terns. The dry, forested hills above the town represent one of the best locations on the island for Caribbean Elaenia and the endemic subspecies of House Wren. Lesser Antillean and Short-tailed Swifts are often drawn to the billowing updraughts synonymous with this mountainous region of the island and can regularly be seen wheeling and soaring over forested areas.

CENTRAL

Des Cartiers Rainforest (13°48′49.5″N 60°57′23.8″W)

In the car park, keep watch around the picnic tables for the highly inquisitive and (if you are eating) often opportunistic Lesser Antillean Bullfinch. If you take the trail from the car park towards the observation area, while en route be on the lookout for endemic St Lucia Oriole in the towering endemic Lansan trees and follow the ethereal call of Rufous-throated Solitaire and shriek of Broad-winged Hawk to their respective locations deep in the heart of the forest. Once at the observation point, look out across the valley for St Lucia Parrots flying by at or below eye level, or feeding in their favoured fruiting trees. The lush vegetation surrounding the viewing area is also a draw for Green-throated and Purple-throated Caribs, Bananaquit, Scaly-naped Pigeon, Grey Trembler, Lesser Antillean Flycatcher and, in the tangled canopy above, Lesser Antillean Euphonia.

BARBADOS

SOUTH

Chancery Lane Wetland (13°03′45.1″N 59°30′00.8″W)

Within the Button Mangroves, Bearded Fig Trees and wispy Casuarinas that broadly encircle this semi-natural wetland can be found a host of resident breeding species. These include 'Golden Warbler', the endemic Barbados Bullfinch, Black-whiskered Vireo, Zenaida Dove, Carib Grackle, Shiny Cowbird, Scaly-naped Pigeon and Caribbean Elaenia. Mudflats and shallow ponds ensure this site is also ideal for migratory North American waders, from

A Green Heron foraging in a shallow tidal pool on the east coast of Barbados.

Willet, Whimbrel, Greater and Lesser Yellowlegs and Short-billed Dowitcher, to smaller waders such as Semipalmated, Spotted, White-rumped and Least Sandpipers. From late September, Belted Kingfisher and species of migrant waterfowl, such as Blue-winged Teal, Ring-necked Duck, Lesser Scaup and American Wigeon start to arrive, adding to the number of aquatic species present year-round (which include Green Heron, Common Gallinule and American Coot). This wetland is primarily rain-dependent and hosts the greatest number of birds between July and November when the pond levels are at heights capable of sustaining the widest variety of species. Note also that the surrounding grasslands are ideal for Grassland Yellow-finch and attract migrant swallows, along with the breeding Caribbean Martin.

Graeme Hall Swamp (13°04′21.0″N 59°34′41.1″W)

The largest expanse of mangrove wetland on the island harbours a wealth of both resident and migrant species. At the time of writing, however, the Graeme Hall Nature Sanctuary is closed, and the remainder of the swamp will prove difficult to explore. Plans for a national park are in discussion, though, and would make the site far more accessible, with wading birds especially standing to benefit from the proposed creation of mudflats within the park. As things stand, a short north–south track runs along the east–west border separating the public lands of the Graeme Hall Swamp and private lands of the Nature Sanctuary. A walk along this can produce migrant herons such as Little Blue and Tricolored, along with resident Cattle, Snowy and

Little Egrets and Green Heron. Other aquatic species at this site include Sora, Common Gallinule and, on occasion, Purple Gallinule. This can also be a rewarding site during autumn migration for sightings of visiting North American warblers, including Northern Waterthrush, Blackpoll and Prothonotary Warblers. This location is also favoured year-round habitat for a significant population of the island's resident 'Golden Warblers'.

EAST

Bayfield Pond (13°10′00.9″N 59°27′09.1″W)

Tucked away in the tiny village of Bayfield in St Philip is a small lily pond that can be particularly rewarding for such breeding residents as Masked Duck and Common Gallinule. Eared Dove, Caribbean Elaenia, Grey Kingbird and Common Ground Dove are regularly seen in the vicinity.

Woodbourne Shorebird Refuge (13°06′06.4″N 59°29′56.5″W)

This former 'shooting swamp' – that is, a site where hunters once targeted arriving flocks of waders – is now maintained primarily to attract and provide a feeding and rest stop for waders during migration. The peak migratory months of July to October can see Woodbourne host an array of species, including Greater and Lesser Yellowlegs, Grey and American Golden Plovers, Short-billed Dowitcher, Pectoral, Solitary and Least Sandpipers, Wilson's Snipe and many more. Also known to frequent this site are Black-crowned Night-heron, Black-bellied Whistling Duck, Green Heron and Common Gallinule.

ST VINCENT

SOUTH AND WEST

Buccament Valley and the **Vermont Nature Trail** (13°12′19.3″N 61°12′45.1″W)

With over 250 species in its flora, this lush, forested valley is a terrific location for sightings of the endemic St Vincent Parrot. The Vermont Nature Trail represents an ideal means of accessing this area. The peak activity times for the parrots tend to be at dawn and dusk, with birds either setting off from, or returning to, their roosts (fewer sightings are logged during the heat of the day). Listen for these vocal parrots calling as they fly between fruiting trees and scan the branches for sightings. A plethora of regional specialties can also be seen here, including Lesser Antillean Bullfinch, Scaly-naped Pigeon, Caribbean Elaenia, Spectacled Thrush, Brown Trembler and Scaly-breasted Thrasher.

NORTH

La Soufrière

Take the La Soufrière Cross-Country Trail, starting at the Rabacca trailhead, for your best chance of sightings of the extremely rare and localised endemic Whistling Warbler. As you ascend the trail listen for the bird's tell-tale and remarkably loud, whistled song. The species is fond of foraging amongst the tangled mosses that cling to the trunks of mature trees, and within stands of heliconia. This lush forest is also home to a host of other species and, as you walk along the trail, take advantage of gaps in the vegetation to enjoy sightings of the near-endemic Grenada Flycatcher, and Lesser Antillean Tanager, Rufous-throated Solitaire, dark morph Bananaquit, all three hummingbird species found on the island, the endemic subspecies of House Wren, Black-whiskered Vireo, Scaly-naped Pigeon and Cocoa Thrush. Be sure to walk slowly, as skittish Ruddy Quail-doves are known to stroll along the trail and will flush on your approach. If you reach a dry riverbed in the heart of the forest (about an hour's walk from the start of the trail), scan the skies above for soaring Common Black and Broad-winged Hawks.

GRENADA

SOUTH

Mt. Hartman National Park (12°00′49.7″N 61°44′46.4″W)

This site is known to harbour the largest remaining population of the Critically Endangered endemic Grenada Dove. A good technique to adopt when attempting to see this species is to slowly wander quietly along the trail network listening for its tell-tale sombre call, while also scanning branches at or below eye level for perched birds. This dove is fond of foraging on the ground, so another option is to sit quietly and wait for it to approach you of its own accord. Patience is needed, but the species is there. If you have time, climb the observation tower and scan the skies for soaring Hook-billed Kites and Grey-rumped and Lesser Antillean Swifts. The mixed vegetation in the vicinity of the tower represents suitable habitat for Grenada Flycatcher, Carib Grackle, Yellow-bellied Elaenia, Antillean Crested Hummingbird, the endemic subspecies of House Wren, Common Ground and Eared Doves and Tropical Mockingbird.

CENTRAL

Grand Etang National Park (12°06′10.9″N 61°41′34.6″W)

This park protects a wonderful expanse of forest. Explore this vast site for prime opportunities to view Lesser Antillean Tanager and Rufous-breasted Hermit, as well as a host of other forest-dwelling species, including Ashy-faced Owl, Broad-winged Hawk, Brown Trembler, Spectacled Thrush, Purple-throated Carib and Ruddy Quail-dove. This is also the only location for Channel-billed Toucan in the Lesser Antilles.

NORTH

Levera National Park (12°13′10.1″N 61°36′39.7″W)

This spectacular 200-ha park exists to conserve an expansive mangrove ecosystem and other prime coastal habitat. The observation tower provides wonderful views of the wetland and a variety of species that includes Cattle and Snowy Egrets, Great Blue Heron, Osprey, Belted Kingfisher, Pied-billed Grebe and Ruddy Duck. Because of its proximity to the coast, species such as Brown Booby, Laughing Gull, Royal Tern and Magnificent Frigatebird can also be observed around the park.

Rufous-vented Chachalaca *Ortalis ruficaudaw* L 57cm

A large, primarily ground-dwelling gamebird that broadly resembles a female Common Pheasant *Phasianus colchicus*. It has a grey head and a long grey neck, brown upperparts, buff underparts and a long, dark, cinnamon-tipped tail. The long legs and short decurved bill are grey and bare red skin shows on the throat.

Vocalisations: loud raspy squawks – much like a chicken in distress

Where to see: introduced to Union Island and Bequia in (St Vincent) and the Grenadines.

Helmeted Guineafowl *Numida meleagris* L 58cm

Unmistakable. A heavy-set dark grey gamebird with white dots and flecks throughout and an obvious 'horn' atop the head. The face and neck show bare blue skin, with much red around the bill area. The short decurved bill and long thick legs are grey. Most often seen on the ground.

Vocalisations: a loud, rapid *tek-uh-tek-uh-tek-uh-tek.*

Where to see: Native to Africa and introduced to Saint Martin and Barbuda.

Red Junglefowl *Gallus gallus* L 25–40cm WS 38–51cm

The instantly familiar and recognisable male has an erect bright red comb atop his crown and a dangling red wattle – it's the ancestor of the domestic chicken. The thickly feathered neck is bright golden-orange, while the breast and belly are a deep, dark green and the back a rich rufous-orange. The smaller female is dark brown above and paler brown below, with red facial skin. The tail feathers are long, elegantly drooped and glossy green in males, but shorter, erect and dark brown in females.

Vocalisations: the male has the typical rooster crow of *cock-a-doodle-doo*.

Where to see: introduced. Domesticated birds are regularly seen in rural communities, while feral populations exist in woodland on Martinique, Guadeloupe and the Grenadines.

♀

♂

Black-bellied Whistling Duck *Dendrocygna autumnalis*

L 47–56cm WS 76–94cm

A tall erect duck with long, pink legs, a bright pink bill and a notable white eye-ring. Adults have a warm grey face, throat, breast and collar, together with a cinnamon-brown crown and neck. The back and wings are a duller brown and the belly is black. The vent is heavily mottled black and white.

Vocalisations: in flight, a high-pitched undulating whistle: *shreee-wee-weee.*

Where to see: found throughout the region and breeds on Barbados. Frequents wetlands, especially those surrounded by trees and dense vegetation.

West Indian Whistling Duck *Dendrocygna arborea*

L 48–58cm WS 85–93cm

A large, tall, erect duck with a chestnut-brown crown, breast, back and wings, along with substantial black-and-white mottled feathering on the belly, flanks and vent. The face is predominantly greyish-beige and, although the throat is white and unmarked, the white front and sides of the neck are finely streaked with black. The bill is black, and its long grey legs extend far beyond the tail in flight.

Vocalisations: a 'creaky' whistled *screee-wee*, uttered in flight.

Where to see: from Dominica northwards, with McKinnon's Saltpond on Antigua a reliable location.

Fulvous Whistling Duck *Dendrocygna bicolor* L 45–53cm WS 85–93cm

A handsomely plumaged, tall, erect duck, with a tawny-coloured head and underparts, a single dark streak on the nape, and a pale, creamy patch on the throat, which is finely streaked with black. The back and wings are a rich dark brown and bright yellowish-beige streaks extend at a 45-degree angle along the flanks.

Vocalisations: a sharp *prch-EEE* can be uttered every second.

Where to see: rare throughout the region, but most often seen on Antiguan wetlands.

Masked Duck *Nomonyx dominicus* L 30–36cm WS 43cm

The smallest duck in the region. Breeding males have strikingly rich rufous-chestnut plumage, which is mottled black on the flanks, back and wings, a black face and crown, a white eye-ring and a dazzling neon blue bill. The lengthy black tail is often held erect. Females have paler beige-brown mottled and streaked plumage, a dark crown and two distinct, thick dark lines running horizontally across a pale beige face. In flight the wing shows an obvious white patch.

Vocalisations: a soft cooing, which is used primarily in courtship.

Where to see: isolated lily ponds on Barbados, Martinique and Guadeloupe. The species seldom flushes into flight, opting instead to submerge until only the bill and head are visible

♂

♀

Ruddy Duck *Oxyura jamaicensis* L 35–43cm WS 58cm

Broadly similar to Masked Duck in size and appearance, with breeding males also showing neon-blue bills and rufous-chestnut plumage. Note, however, the large white cheeks and prominent black cap and nape of the male Ruddy. The females are dark brown above and paler beige below with significant light and dark barring throughout and show a dark-brown cap that descends to eye level, with a single dark horizontal line across the pale face. No white patch is visible in the wing in flight. Largely silent.

Where to see: rare, with Anguilla, St Kitts, Saint Martin and Grenada producing the most sightings. Similar habitat to Masked Duck.

♂

♀

♂ non-br.

Muscovy Duck *Cairina moschata* L 64–77cm WS 140–150cm

A very large, darkly plumaged duck with a black body and dark green, iridescent wings. Red, globular, wart-like knobs of skin surround the eye and bill. In flight, a significant amount of white shows in the wing. Largely silent.

Where to see: the smaller domesticated version of the Muscovy Duck has numerous colour variations and is common in rural communities throughout the region. Non-domesticated, 'authentic' birds can still be found in some wetlands and large waterbodies on Martinique.

Ring-necked Duck *Aythya collaris* L 41cm WS 68cm

♂

♀

The handsome male has a jet-black head, neck, breast, vent and upperparts, all of which contrast markedly with its snow-white belly and pale-grey vermiculated flanks. There is a white finger-like patch at the fold of the wing. The eye is a rich, golden-orange. Females show a greyish-beige face with a white eye-ring, a dark-brown crown and upperparts, and paler beige underparts punctuated by a white belly. Both sexes show a white ring encircling the grey bill towards the black tip (males also have a white band at the bill base) and pointed crowns. Mostly silent.

Where to see: a common autumn migrant that prefers wetlands and large bodies of freshwater.

Greater Scaup *Aythya marila* L 46cm WS 78cm

A large, heavy-bodied duck with a dark green, rounded head and neck, and a white belly and flanks sandwiched between a black breast and vent. The back shows fine black and white vermiculations. The broad greyish-blue bill is tipped black, and the eye is bright yellow. Females are dark brown (darker on the head, breast and upperparts, and paler below) and can be distinguished from females of most other waterfowl species by the thick white patch at the base of the bill. Largely silent.

Where to see: a very rare migrant, with the occasional bird seen on coastal wetlands.

Lesser Scaup *Aythya affinis* L 43cm WS 71cm

The sexes are almost identical to those of Greater Scaup, but this is a smaller duck with a notably peaked rear to its crown and a narrower greyish-blue bill. The male's head and neck have a deep purple sheen in the right light. Mostly silent.

Where to see: a regular migrant that is often seen in small flocks on wetlands near to the coast, or occasionally on larger waterbodies further inland

Northern Shoveler *Spatula clypeata* L 50cm WS 78cm

A long-bodied duck with a large, flat, spatula-shaped bill (this is black in males and yellowish-brown in females). Breeding males show a dark emerald-green head and neck, a white breast and a chestnut-orange belly and flanks. The back and vent are largely black. Females have greyish-brown plumage (darker on the body and paler on the face) with clearly demarcated, pale-edged, individual feathering on the back and wing. Largely silent.

Where to see: a scarce migrant to the region. This species often appears singly or in pairs in mangrove swamps and other wetlands.

Blue-winged Teal *Spatula discors* L 38cm WS 62cm

Breeding male shows glossy chestnut-brown breast, flanks and belly – all of which are heavily speckled black. Upperparts are a dark, greyish-brown. The head is a rich greyish-purple with a large white crescent at the base of the black bill. The female is mostly ash-brown with clearly demarcated and pale-edged individual feathering giving a scaly look to the body overall. She has a pale split eye-ring, a dark eye-stripe and a large white spot at the base of her bill. In flight, both sexes show a large bright blue patch in the wing. Usually silent.

Where to see: this is a common autumn migrant from North America seen on waterbodies of varying sizes.

American Wigeon *Mareca americana* L 52cm WS 76–92cm

A large, heavy-set duck in which the breeding male has milk chocolate-coloured upperparts with fine, dark vermiculations on the back and wing, pinkish-brown on the breast and flanks, a cream-white belly and a black vent. The crown and forehead are pale and unmarked, but the remainder of the head is speckled black. A broad, bright green band descends from the eye onto the neck-sides. The head of the female is a light greyish-brown with dark speckling, while the upperparts are dark brown, and the breast and flanks a soft rufous-orange. The belly is cream. Generally silent.

Where to see: a regular autumn migrant often seen in pairs or small flocks on fresh and brackish waterbodies.

White-cheeked Pintail *Anas bahamensis* L 38–50cm WS 55–65cm

A small, somewhat bulbous-headed duck, with pale beige-brown underparts and flanks heavily speckled with dark chocolate. The crown and upperparts are a rich dark brown. Key features include large white cheeks and a white throat, along with a two-toned bill of reddish-pink and dark grey. The long, very pale tail is thin and sharply pointed.

Where to see: breeds on islands north of Guadeloupe, where it can be found on small, isolated ponds; also, in larger numbers on more extensive waterbodies.

Northern Pintail *Anas acuta* L 57–58cm WS 88cm

The breeding male is largely grey and white with darkly streaked upperparts. It has a warm brown head with a single vertical white stripe on chocolate-brown neck-sides and a black-and-silver bill. Two long, thin, stiff, black tail feathers protrude from its sharply pointed tail. The female has a pale brown unmarked face, dark brown upperparts and greyish-beige underparts. Clearly demarcated individual feathering creates a heavily chevroned appearance to the back, breast and flanks.

Where to see: a rare migrant, with wetlands on Guadeloupe and Barbados the best for sightings.

Pied-billed Grebe *Podilymbus podiceps* L 34cm WS 60cm

A chunky, rather drab, greyish-brown aquatic bird, with a lengthy, stout neck and short, thick, pale grey bill. Breeding birds show an obvious black band encircling the bill, as well as a black throat patch which fades or disappears in non-breeding birds. The juveniles show faint black and white streaking on the head and neck.

Vocalisations: an accelerating series of *hup* calls leading into *how-wu* sounds and occasionally seal-like *aarrps*.

Where to see: regular on inland waterbodies.

Red-billed Tropicbird *Phaethon aethereus*

L 90–105cm including tail-streamers at 46–56cm WS 99–106cm

One of the most stunning seabirds on the planet. The plumage is snow-white with an intricate pattern of black horizontal lines drawn across the back, rump and some of the upperwing. Wing-tips are heavily 'inked'. Two exquisitely painted black streaks extend upwards from the base of the bright red bill, through the eye and towards the crown. Long, white tail-streamers extend behind the bird in flight. The juvenile has a yellow bill and no tail-streamers.

Vocalisations: harsh, tern-like screams at the breeding colony: *skreeee–skreee–skreee*.

Where to see: possible at sea throughout the region, but the cays north of Anguilla, and Cape Moule a Chique on St Lucia offer fabulous views.

White-tailed Tropicbird *Phaethon lepturus*

L 71–80cm about half of which includes its tail-streamers WS 89–96cm

Smaller and less robust than Red-billed Tropicbird, with adults showing a diagnostic thick black V on the upperside, widest on the upperwings and tapering to its narrowest point at the rump. The tail carries two long creamy-white streamers. A black streak extends from the base of the bright orange bill through the eye. Wing-tips are heavily 'inked'. Juveniles have yellow bills, black barring on their wings and backs, and lack tail-streamers.

juv.

Vocalisations: poorly known but described as a harsh, rasping shriek.

Where to see: possible at sea throughout the region; breeds on Sint Maarten (Cupecoy), Saint Martin (Precipice des Oiseaux) and Dominica (Ponte Michel, Canefield and Tarou Cliffs).

Rock Dove *Columba livia*
L 28–33cm WS 50–56cm

Its overall plumage pattern and colour varies, but the typical, ancestral-type Rock Dove has a largely grey body, darker bluish-grey head and breast, a neck with a purplish-green iridescent sheen, two broad black wing-bars and dark wing-tips. The bill is black with a white cere, and the legs are red. Juveniles are duller overall with reduced iridescence.

Vocalisations: the familiar town-pigeon cooing.

Where to see: possible everywhere from coastal sea cliffs to towns and bustling city centres.

Eurasian Collared Dove *Streptopelia decaocto* L 30–33cm WS 45–55cm

Longer and paler than any other dove in the region, with light creamish-beige plumage, a distinct semi-circular black ring on the neck (not joined at the front), a deep red eye and a white orbital ring. The square tail is broadly tipped with white. Juveniles do not show the neck ring and the eye is brown.

Vocalisations: a repetitive and low *whoo–whoo …whoo–whoo.*

Where to see: often seen in residential neighbourhoods and hotel grounds, but possible in all habitats.

White-crowned Pigeon *Patagioenas leucocephala*

L 29–35cm WS 48–59cm

Similar in size and overall slate-grey plumage to the regionally more widespread Scaly-naped Pigeon, but with a brilliant snow-white cap (greyish-white in females) and a pale-yellow eye. Green iridescence shows well on the heavily scaled nape and neck-sides. The legs are pink, and the bill is two-toned with pink at the base and cream at the tip. Juveniles are greyish-brown with a pale forehead.

Vocalisations: a high-pitched *hwhOOOO-hoo-hoo-whooo … hoo-hoo-whooo …*

Where to see: occurs north from Guadeloupe. Prefers mangrove forest and mixed woodland, but also frequents hotel grounds and areas of parks and capital cities with trees.

Scaly-naped Pigeon *Patagioenas squamosa* L 32–41cm WS 58–66cm

The largest pigeon in the Lesser Antilles is a dark, slate-grey with a deep purple head, throat, foreneck and upper breast. The nape and neck-sides are heavily scaled and show a rich purple, metallic sheen. The brilliant red eye is broadly framed by orange-peach orbital skin. The legs are pink and the bill two-toned, with dark purple at the base and a cream tip. The juvenile is slate grey with a dull brownish head and neck.

Vocalisations: a rolling *HOOOO-hooo-hoo-HOOOO-hooo-hoo-HOOOO.*

Where to see: forested areas throughout the region. Common on Barbados, where it regularly frequents hotel grounds and residential gardens.

Ruddy Quail-dove *Geotrygon montana* L 19–28cm WS 40–41cm

A short, chunky dove with dark, rich rufous-brown upperparts, often tinged with reddish-purple iridescence, and paler tawny-buff underparts. Note the obvious pale throat, the cream patch at the bend of the folded wing and the broad, creamish-buff, horizontal streak underneath the eye. Females and juveniles show dark greyish-olive upperparts and paler beige underparts, and have duller, less obvious, facial features.

Vocalisations: a plaintive *whHOOooo*, repeated roughly every four seconds.

Where to see: a secretive forest dweller most numerous on St Lucia, St Vincent, Guadeloupe, Martinique and Dominica, often seen along forest trails.

♂

♀

juv.

Bridled Quail-dove *Geotrygon mystacea* L 27–30cm WS 48cm

Adults have dark, rich, greyish-brown heads and upperparts, pale throats, rufous-pink-tinged breasts and creamish-buff bellies. The dark, greyish-brown wings are highlighted by chestnut wing-tips and, in good light, green and purple iridescence shows on the nape, neck-sides and upper back (this is absent in juveniles). Also, note the thick white horizontal streak underneath each eye. The bill is two-toned (pink at the base and cream at the tip).

Vocalisations: a long, low, drawn-out *WHOOOOOO.*

Where to see: a secretive forest dweller found on most islands in the north of the region as far south as St Lucia.

Grenada Dove *Leptotila wellsi* L 28–31cm WS 50cm

A subtly patterned dove characterised by rich brown (almost auburn) upperparts and wings, a buff-coloured breast (tinged pink in good light) and a creamy-white belly. The face is a pale pinkish-cream, and the forehead and crown are a soft grey (darker on the latter). A vertical white streak shows at the bend of the folded wing. The juvenile is duller overall with a tawny hue to its upper back and breast.

Vocalisations: a soft and plaintive *hooOOoo* repeated roughly every eight seconds.

Where to see: endemic to Grenada, with the largest numbers found in Mt Hartman National Park.

White-winged Dove *Zenaida asiatica* L 27–29cm WS 48-58cm

Greyish-brown dove with an obvious long white streak running along the lower edge of the folded wing. A purple-and-gold metallic sheen shows on the neck-sides and vibrant blue skin surrounds the amber eye. The bill is long, thin and black, and the legs are a deep pink. The juvenile is pale grey overall and lacks the metallic sheen.

Vocalisations: a coarse, 'barked', almost rooster-like *aroooh-roo-hoo-hoo-hoo.*

Where to see: mostly found on Anguilla, Antigua and Barbuda, and other northern islands, where it forages on the ground among dry scrubby vegetation.

Zenaida Dove *Zenaida aurita* L 30–31 cm WS 36–41 cm

A common rufous-cinnamon dove with a bright, metallic, purple sheen to its neck-sides. The breast is rich chestnut and can appear rose-tinted, while the belly is a soft greyish-cream. Black splotches show on the folded wing and the wing-tips are fully black. A short, black streak is present at the base of the cheek with another occasionally seen behind the eye. Juveniles are duller and lack purple sheen to neck-sides.

Vocalisations: a haunting *whoo-hoOO-hoo-hoo-hoo.*

Where to see: common in all habitats. Regularly frequents areas of human habitation, foraging on the ground.

Eared Dove *Zenaida auriculata*

L 22–28cm WS 13–17cm

Slightly smaller and slimmer than the similar Zenaida Dove, with a paler beige-brown plumage that is notably rose-tinged on the head, neck and breast. At rest, the tail is significantly more pointed and 'pintail-like' and, in flight, the outer-tail feathers show chestnut tips. Note the pale blue eye-ring and golden metallic sheen on the neck-sides. Juveniles are brownish-beige, with golden flecks on the neck, breast and wings.

Vocalisations: a hoarse *hrooo–horrgh–hrooo–horrgh.*

Where to see: common in Grenada and St Vincent, but less so in Barbados. Prefers dry forest and scrub.

Common Ground Dove

Columbina passerina

L 16cm WS 22–25cm

The tiniest dove in the Caribbean, with soft, 'dusty', greyish-brown plumage, and many pinkish-silver scales covering its crown, head, neck and breast. Its rich chestnut-coloured wing-tips and underwings are clearly visible in flight. The upperwing is decorated with random black splotching. The legs are soft pink.

Vocalisations: a series of rapid-fire coos (*whooo–whooo–whooo*).

Where to see: dry scrub and grasslands, quiet roadside verges and residential neighbourhoods.

Common Nighthawk *Chordeiles minor* L 21–25cm WS 53–57cm

In flight, shows a lengthy cigar-shaped body, long, slender, pointed wings and a forked tail. The head appears somewhat flattened, the eyes large and black, and the bill is disproportionately small relative to its body size. The upperparts are mottled black, brown, grey and beige. The cream underparts are heavily barred. A broad white wing-patch just before the wing-tip is obvious in flight. Males also show a broad, white, partial tail-band and a white throat (buff in females).

Vocalisations: a nasal *veent*.

Where to see: typically, only seen in region during spring and autumn migration. Mostly crepuscular and frequents open xeric landscapes and coastal sites. Roosts on bare branches, occasionally on the ground.

Antillean Nighthawk *Chordeiles gundlachi* L 20–22cm WS 50–55cm

Almost identical to Common Nighthawk, but a little smaller and with plumage colouration and contrast more marked. The two species are best differentiated by voice.

Vocalisations: a rapid *preet-ah-preet-ah-preet*.

Where to see: typically, only present in the region during the summer months, where it can usually be seen in dry open landscapes such as cane fields and barren agricultural areas, grassland and scrub. Although primarily crepuscular, it occasionally flies during the day especially during overcast conditions.

White-tailed Nightjar *Hydropsalis cayennensis* L 21–23cm WS 53–57cm

A small compact nightjar with a relatively large flat head, buff eyebrow and broad chestnut collar. The male's upperparts are mottled chestnut, grey, black and beige. The underparts are white with faded buff and black chevrons (absent on the unmarked white underside of tail). In flight, males show a white patch just before the wing-tips (this is chestnut in females, which also show significantly less white throughout the body). Feeds on insects in open grassland by launching itself up from the ground.

Vocalisations: a high drawn-out *chip-phweeeee*.

Where to see: Martinique, where it roosts on the ground by day under low-lying vegetation.

Rufous Nightjar (St Lucia Nightjar)

Antrostomus rufus L 25–30cm WS 57–63cm

Large nightjar with dark rufous upperparts, heavily interspersed with paler beige splotches and black streaking. The underparts are a slightly paler reddish-brown and significantly dappled. The large head is flat, with an obvious thin white throat band. In flight, there is no white or chestnut band in the wing (males do, however, have white-tipped tails).

Vocalisations: a distinctive four-noted *TIU … weh–weh–WEOU*. Males call vociferously at dusk and can often be traced to their perch.

Where to see: considered a full endemic species (St Lucia Nightjar) by some, and a subspecies of Rufous Nightjar by others. Primarily found in the dry woodlands and grasslands of north-east St Lucia. Best observed during the February–April breeding season.

Black Swift *Cypseloides niger*
L 20cm WS 45–46cm

A large, totally black unmarked swift with a subtly notched tail and long sickle-shaped wings.

Vocalisations: a long *trrrrriiiiiip* sandwiched between shorter *tip-tip* notes.

Where to see: found throughout the region between March and September, nesting on sea cliffs along the south-eastern Barbados coastline. Often seen over the high mountainous spine of St Lucia and other highland forest regions.

White-collared Swift *Streptoprocne zonaris* L 21cm WS 50cm

A large, black swift, the males of which show an obvious broad white collar completely encircling the neck. This collar is narrow or broken in females and faded in juveniles; note that Black Swift is always completely unmarked. The wings of Collared are also more angled and sickle-shaped than those of Black Swift, while the tail is more obviously forked.

Vocalisations: short, conjoined wheezy whistles.

Where to see: feeds in large flocks at great height on Grenada.

Lesser Antillean Swift *Chaetura martinica* L 10–12cm WS 28cm

A small, rather drab, brown swift, with a wide square tail. The darker brown upperparts contrast with paler brown underparts, while the throat is paler still. Note the pale band across the rump.

Vocalisations: a series of very high-pitched twittering whistles – *pttttreeeeeee* and *preep* notes.

Where to see: endemic to the Lesser Antilles, breeding on Guadeloupe, St Lucia, St Vincent, Martinique and Dominica. Flocks are often seen feeding high over moist tropical forest.

Grey-rumped Swift *Chaetura cinereiventris* L 10–12cm WS 28cm

The head and upperparts of this small, slender swift are a shiny dark blue (often appearing glossy black at a distance). A large, pale grey patch is present on the rump. The underparts are primarily medium grey, with dark bluish-black vent and underside of the short, square tail.

Vocalisations: a series of twittering *preep* and flatter rolled notes.

Where to see: commonest on Grenada, where it occurs over forests.

Short-tailed Swift *Chaetura brachyura* L 10–12cm WS 28cm

Similar in size to Lesser Antillean Swift but has glossy bluish-black upperparts, dark underparts and a pale rump and vent. The minuscule tail barely protrudes beyond the rump, giving this species the appearance of a small, long-winged, short-bodied bat.

Vocalisations: a mix of single *chup* and *tzsiip* notes.

Where to see: breeds on St Vincent, wandering to St Lucia and Grenada. Primarily feeds above areas of secondary forest, mixed woodland and scrub, often alongside Lesser Antillean Swift on St Lucia.

Rufous-breasted Hermit *Glaucis hirsutus* L 11–12cm WS 11cm

Distinctive among hummingbirds in the region due to its rusty-orange breast, this medium-sized hermit shows dark green upperparts, chestnut panelling in its white-tipped tail and a long, decurved bill, darker above and a pale yellow below.

Vocalisations: a rapid-fire high-pitched *peihp-peihp-peihp … peee-pee-pe* tailing off towards the end.

Where to see: generally at higher elevations on Grenada, especially in nutmeg and cacao plantations.

Green-throated Carib *Eulampis holosericeus* L 11cm WS 11cm

A relatively large hummingbird with a brilliant, neon-green throat and a dazzling sapphire-blue crescent across its breast. The head and upperparts are a metallic bronzed green, interspersed with shards of blue. Blue is also present on the rump and vent. The belly is black, as are the relatively long wings and tail. The lengthy black bill is decurved (more so in females).

Vocalisations: a very rapid flat *tihck-tihck-tihck*.

Where to see: most well-vegetated habitats in the region, from forests to residential gardens. A frequent visitor to the bright orange flowers of *Cordia sebestena*.

Purple-throated Carib *Eulampis jugularis* L 11–12cm WS 12cm

The largest hummingbird of the region. It has a shimmering heavily scaled pinkish-purple throat and breast, perfectly framed by a jet-black head, nape, back and belly. The wings are metallic turquoise-green with a hint of gold, and the tail is glossy blue. The long, black bill is decurved (almost sickle-shaped in females). Juveniles show orange throats and breasts, flecked with red.

Vocalisations: *tsp … tsp … tsp.*

Where to see: a Lesser Antillean endemic known to most islands. Prefers primary forest at high elevations, but possible in most habitats.

Blue-headed Hummingbird *Riccordia bicolor* L 9cm WS 10cm

The male is unmistakable with his striking sapphire-blue head and throat, shimmering turquoise-green and heavily scaled back and underparts, and long, deeply forked, blue tail. The female closely resembles female Antillean Crested Hummingbird, with creamy-white underparts and metallic-green upperparts. Note that the latter has a shorter, more rounded tail, while Blue-headed has a tail that is longer, forked and blue-tipped. The long and straight bill is black (with a red-tinged lower mandible in males).

Vocalisations: *clik-clik-clik.*

Where to see: a near-endemic known only to Dominica and Martinique, where it inhabits moist forests at high elevations.

♂

♂

Antillean Crested Hummingbird *Orthorhyncus cristatus*

L 8cm WS 12–13cm

The smallest and most frequently seen hummingbird in the region. Males have a permanently erect neon-green crest, tipped with purplish-blue in some subspecies. The upperparts are a metallic, bronzed green overall, broken with shards of bright green iridescence. The underparts are brownish-black, save for the pale beige throat. The female has pale beige underparts and soft metallic green upperparts and head (with a slight crest). The tips of her tail feathers are elegantly and individually tipped with white.

Vocalisations: a rapid *tck-tck-tck*.

Where to see: found throughout the region and possible in any habitat.

Smooth-billed Ani *Crotophaga ani* L 34 cm WS 43–45cm

A bizarrely shaped black bill shows a swollen convex bump atop the maxilla. The plumage is glossy black and highlighted by well-defined scaly feathering running from the crown onto the upper back, as well as on the front of the neck and breast. The tail is very long.

Vocalisations: a high-pitched squeal reminiscent of the name – *AHHHH-NII*.

Where to see: found throughout the region, but especially common on St Vincent and Grenada. Often in pockets of disturbed vegetation along roadsides and near to human habitation.

Yellow-billed Cuckoo *Coccyzus americanus* L 26–32 cm WS 43cm

Slightly smaller and slimmer than the similar Mangrove Cuckoo, with medium brown rather than grey-brown upperparts, chestnut wings and snow-white underparts. Note the prominent yellow eye-ring and a heavy presence of yellow on the slightly decurved bill. A bold black-and-white pattern is present on underside of long tail.

Vocalisations: like Mangrove but higher pitched.

Where to see: a regular migrant, often seen alone in mixed woodland.

Mangrove Cuckoo *Coccyzus minor* L 28–33cm WS 38–43cm

A medium-sized cuckoo with characteristically bold black-and-white patterning on the underside of the very long, broad tail. The upperparts are typically warm greyish-brown and the underparts buff yellowish-cinnamon. A dark cap descends to the eye and sits above pale cheeks, while a dark eye-stripe creates a masked appearance. The decurved bill has a dark grey maxilla and a yellow mandible. Juveniles are duller overall.

Vocalisations: a frog-like *koh-koh-koh-koh.*

Where to see: found in a range of habitats on most islands, from dense tropical forest in Dominica to dry scrubland on Barbuda. Absent from Barbados.

Clapper Rail *Rallus crepitans* L 37cm WS 48–51cm

A gallinule-sized, greyish-brown bird with a long neck, prominent cream eyebrow, and a lengthy, slightly decurved, orange-yellow bill. The upperparts are heavily streaked with charcoal grey, while the plain, greyish-brown underparts have fine white barring on the belly and flanks. The short tail is often held cocked and the legs are orange-yellow.

Vocalisations: a series of increasingly rapid *kick-kick-kack* notes.

Where to see: wetlands on Barbuda and St Kitts, where crepuscular.

Sora *Porzana carolina* L 22cm WS 30–31cm

A small, plump, compact rail with medium-brown upperparts streaked with longitudinal lines of black and white splotches. The grey face shows a broad black mask covering the eye and a short, sharply pointed, yellow bill. Black on the throat can bleed downwards to the central breast area, while the remaining underparts are grey with thin white and black barring on the belly. A short, pointed tail shows white on the underside and is often held erect.

Vocalisations: an increasingly rapid, repeated *er-WHEET*.

Where to see: a common autumn migrant that frequents wetlands with dense surface vegetation, surrounded by thick sedge.

Purple Gallinule *Porphyrio martinica* L 27–36cm WS 50–55cm

Broadly resembles Common Gallinule, but significantly more colourful, with a shimmering indigo-blue head, neck, breast and belly, and a bronze-green back and wings. The brilliant bright red bill is tipped with yellow, and the pale frontal shield stands out. The legs are long and yellow. Juveniles are brown with a soft greenish-bronze hue to the back and wings.

Vocalisations: a repetitive *tengk tengk tengk tengk*.

Where to see: a migrant throughout the region, but also a breeding resident on Montserrat, Guadeloupe and Martinique. Prefers wetlands with thick surface vegetation.

Common Moorhen (Common Gallinule) *Gallinula galeata*

L 34cm WS 54–62cm

An aquatic species with rich, sooty-black plumage, prominent bright-red bill and frontal shield, thin yellow legs with red 'garters' and long toes. The slender juveniles show brownish-grey plumage with greenish-yellow legs and lack the red frontal shield. When alarmed, flicks its tail forwards and backwards, flashing the white outer tail feathers.

Vocalisations: a soft, repeated *kuhp* … *kuhp* usually accompanied by tail flicking. Also, *tuh-kuh-tuk-tuk-tuk-tuk-tuk-tuhhh*.

Where to see: common throughout region on waterbodies of all sizes.

American Coot *Fulica americana* L 37–38cm WS 58–71cm

An aquatic species with a charcoal-grey body and a black neck and head. Can be distinguished from the similar Common Gallinule by its larger, bulkier build, white bill and frontal shield, and thicker, duller yellow legs. Juveniles show an olive-brown crown and upperparts, pale grey underparts and a greyish bill.

Vocalisations: a variety of cackling clucks and croaks.

Where to see: found on wetlands throughout the region, breeding on several islands including Barbados and St Lucia.

Wilson's Storm-petrel *Oceanites oceanicus* L 18cm WS 37cm

A small pelagic species with a soft, graceful flight, sometimes appearing to 'dance' on the surface of the ocean, with wings held aloft, dipping its yellow-webbed toes delicately into the water. Its plumage is dark brown and there is a broad, U-shaped white band on the rump which descends onto the underparts. The wings are rounded, the tail square. The bill is short and black, and the black legs extend beyond the tail in flight. Largely silent.

Where to see: from late spring through summer, the deep ocean waters off Guadeloupe can produce regular sightings, but it is possible at sea throughout.

Leach's Storm-petrel *Oceanodroma leucorhoa* L 21cm WS 47cm

Very similar to Wilson's Storm-petrel, but a larger bird, with slightly paler plumage, an obviously forked tail and longer, pointed wings. There is no yellow webbing on the toes, and the legs do not extend beyond the tail in flight. Unlike Wilson's, the U-shaped white band is restricted to the rump and is often split by a dark central line. Largely silent at sea.

Where to see: possible throughout, but reliable off Guadeloupe in spring and Barbados in autumn.

Black-capped Petrel *Pterodroma hasitata* L 41cm WS 101cm

A medium-sized seabird with brownish-black upperparts punctuated by a thick white collar and a broad white 'U' extending from the rump onto the lower back. A dark cap covers the eye, and the forehead is white. The upperwing is brownish-black, while the underwing is white and dark-edged, as is the underside of the tail. The underparts are white. The species has an elevated and buoyant flight, yielding to the winds on bended wings. Silent at sea.

Where to see: highly pelagic with most sightings around northern isles. Likely nests in remote mountainous regions on Dominica.

Great Shearwater *Ardenna gravis* L 48cm WS 109cm

Almost identical in plumage and pattern to Black-capped Petrel but larger, with a dark forehead, dark belly and a significantly narrower white collar and U-shaped band on the rump. The bill is also longer and slenderer. Tends to glide low over the water on stiff wings. Silent at sea.

Where to see: possible at sea throughout the region during summer, with the best chance from shore on Guadeloupe.

Sooty Shearwater *Ardenna grisea* L 45cm WS 102cm

A stocky seabird with a long, thin, straight bill, hooked at the tip, and a tendency to fly low over the ocean surface on stiff wings. Distinguished from other shearwaters seen in the Lesser Antilles by its completely dark, sooty-brown plumage (the breast and belly are white on all other shearwaters in the region). Silent at sea.

Where to see: rare in the region, but best seen at sea from May to August.

Cory's Shearwater

Calonectris diomedea
L 51cm WS 119cm

Easily distinguished from other shearwaters in the region by its larger size, paler greyish-beige face, crown and upperparts, and its large, pale yellow bill (all others in the region have black bills). The underparts and underwings are white, the latter fringed with brown. Shows a more languid flight style. Silent at sea.

Where to see: possible throughout the region, with May to July being the best time.

Manx Shearwater

Puffinus puffinus L 34cm WS 83cm

Similar to Audubon's Shearwater, but with blackish – not brown – upperparts. Also slightly larger, appearing darker faced, longer winged and shorter tailed. Note the significant amount of white under the tail. The flight style is less 'flappy' than Audubon's. Silent at sea.

Where to see: commonly seen from shore throughout the year, especially off Guadeloupe.

Audubon's Shearwater *Puffinus lherminieri* L 30cm WS 68cm

Relatively short, rounded wings and a long tail. The plumage is almost perfectly halved, with dark brown upperparts and white underparts, though it has a dark undertail. The split coloration continues onto the head, with a dark cap down to the eye and white cheeks. The underwing is white, fringed with brown. Silent at sea.

Where to see: common at sea throughout the region. Breeds in colonies on Antigua and Barbuda, Barbados, Martinique, Grenada and Guadeloupe.

Glossy Ibis *Plegadis falcinellus* L 57cm WS 90cm

An unmistakable egret-sized bird with dark rufous-maroon plumage, a long neck and a lengthy, decurved, greyish-brown bill. In favourable light, the wings and tail show a bronzed, metallic green sheen. The head and neck of non-breeders are finely streaked with greyish white.

Vocalisations: a croaky guttural squawk.

Where to see: uncommon, but possible in wetlands throughout the region.

Least Bittern *Ixobrychus exilis* L 31cm WS 43cm

A small, smartly patterned heron with primarily chestnut plumage, but with a black (dark brown in females), crown, back, rump and tail. The underparts are white but show broad chestnut streaks on the foreneck and breast. The long sharp bill and legs are yellow. The similar Green Heron (page 75) is larger and much darker overall, with a deep rufous rather than chestnut neck.

Vocalisations: a chuckling guttural *auh-auh-auh-auh*.

Where to see: heavily vegetated wetlands on Dominica, Guadeloupe and Martinique.

♀

♂

Black-crowned Night-heron *Nycticorax nycticorax* L 60cm WS 109cm

Roughly the height of Little Egret (page 83) but much 'chunkier', this often-hunched heron has a prominent black crown and back contrasting well against plain grey wings and pale greyish-white underparts. A red eye stands out against a pale face. Thin, white, lanceolate plumes descend from the back of the head. Juveniles are brown with white flecks and streaking. The bill is dark above and pale below.

Vocalisations: sometimes utters a single honking note.

Where to see: a migrant to the entire region, but also resident on Barbados, Grenada, Guadeloupe, Martinique and Antigua. Prefers mangrove swamps and wetlands with significant tree cover.

juv.

Yellow-crowned Night-heron *Nyctanassa violacea* L 60cm WS 107cm

A largely grey heron, with a typically erect stance, black head and the heavy presence of black feathering on the back and wings. A broad yellow streak shows on the centre of the otherwise black crown, and thin white lanceolate plumes protrude from the back of the head. Has a thick white band under the amber-orange eye. Juveniles are brown with white flecks and streaking and heavy all-black bills.

Vocalisations: a harsh throaty *auwk*.

Where to see: a migrant and breeding resident throughout. Primarily crepuscular but can be active at any time. Typically frequents mangrove swamps and rugged coastlines.

juv.

Green Heron *Butorides virescens* L 45cm WS 56cm

A small, stocky heron with a dark crown, rich rufous-maroon head, neck and breast, with the foreneck and breast heavily streaked with white. It has a dark blue-green back and wings showing obvious pale-edged feathers. The belly and vent are grey, while the long bill is yellow and black. Often stands motionless at water's edge or perched on a low branch. Juveniles are rufous brown with heavy white streaking.

Vocalisations: a sharp *KYIUCK … KYIUCK* plus *keh-keh-keh-keh-keh-keh* on landing

Where to see: common breeding resident found in wetlands across the region.

juv.

Cattle Egret *Bubulcus ibis* L 51cm WS 92cm

A medium-sized, all-white heron. Non-breeders show a yellow bill and darkish green or 'dirty yellow' legs, becoming a bright orangey-yellow as the birds enter breeding condition. The crown, breast and back show lengthy buff-orange plumes when breeding, too. Juveniles have dark bills and black legs.

Vocalisations: a throaty *huhhk-huhhk-huhhk*.

Where to see: very common in grasslands and agricultural fields.

br.

non-br.

peak br. condition

Great Blue Heron *Ardea herodias* L 115cm WS 180cm

The largest and tallest heron in the region, with a long, powerful, greyish-brown neck, and gargantuan wingspan. The plumage is a warm blue-grey, with adults showing a pale crown, white chin, streaked foreneck and rufous thighs. The face and head are typically paler than the body, with a dark blue streak over the eye. The juvenile shows a dark crown along with untidy streaking on the foreneck, breast and belly.

Vocalisations: a short, gruff, loud *AUGHK*.

Where to see: a regular migrant in low numbers. Prefers tidal pools, mangrove swamps, freshwater ponds and rivers.

br.

imm.

Grey Heron *Ardea cinerea* L 94cm WS 185cm

Very similar to Great Blue Heron, but slightly smaller, with paler grey plumage and more white in the neck and head. The thighs are white or pale grey (Great Blue has rufous thighs). The chunky bill is yellow.

Vocalisations: a loud *gwark*, given in flight.

Where to see: a rare transatlantic visitor in winter, most frequently seen on Barbados in mangrove swamps and other wetlands.

Purple Heron *Ardea purpurea* L 84cm WS 135cm

A large but significantly shorter and slimmer-bodied heron than Great Blue. The crown is dark, and the long, thin head and neck are chestnut, with distinctive and elegant black streaking. The remaining plumage is slate-grey with shades of brownish-purple, and the belly is black.

Vocalisations: a higher pitched *frank* call than other large herons.

Where to see: a rare transatlantic visitor seen occasionally on Barbados and other southern islands. Prefers wetlands.

Great Egret *Ardea alba* L 97cm WS 155cm

Large, all-white heron with a notably long neck, large yellow bill and long black legs. Breeding birds have spectacular white 'aigrette' plumes – often held erect from their backs – and show stunning emerald-green lores. Non-breeders lack plumes and have yellow lores.

Vocalisations: a grating *auuurrrghk.*

Where to see: resident on most islands from Guadeloupe north, but also a regular migrant throughout to wetlands, coasts and saturated grasslands.

Tricolored Heron *Egretta tricolor* L 63cm WS 90cm

Slender, medium-sized, bluish-grey heron with a white breast, belly and vent. The neck is very long and thin, and the white foreneck shows a broad chestnut streak down its centre. The long, thin, black-tipped bill is blue in breeding birds and yellow in non-breeders. The juveniles are a rusty chestnut-grey with white underparts.

Vocalisations: a honking *auuukgh*.

Where to see: a regular, though scattered, migrant to the region. Prefers wetlands, especially mangrove swamps.

br.

non-br.

juv.

Little Blue Heron *Egretta caerulea* L 63cm WS 100cm

Beautiful medium-sized heron with a dark, slate-blue body, deep purple neck and head and two-toned bill (that is pale grey basally and black at the tip). First- and second-year birds show varied white-and-blue mottling. The juveniles have all-white plumage, a two-toned bill and pale greyish-green legs.

Vocalisations: a raspy *aaaach*.

Where to see: a regular migrant that prefers wetlands and coasts.

Snowy Egret *Egretta thula* L 58cm WS 88cm

Almost identical to Little Egret (opposite), but marginally smaller with a shorter black bill drooping very slightly towards the tip. Breeding birds have a bushy crest of short plumes at the back of the head and white aigrettes. Bright reddish-orange lores are present at the height of breeding (otherwise yellow).

Vocalisations: strangled and raspy croaks and squawks.

Where to see: throughout the region, frequenting wetlands, remote beaches and tidal pools.

br.

non-br.

Little Egret *Egretta garzetta* L 60cm WS 95cm

A medium-sized white heron with a long, straight, black bill and slender black legs with yellow toes. The lores are typically greyish-turquoise, often showing reddish-orange when breeding. Breeders have two long, thin, lanceolate, white plumes protruding from the back of the head, and white aigrettes. Juveniles can show green-tinged black legs.

Vocalisations: strangled-sounding croaks and squawks.

Where to see: mangrove wetlands, tidal pools and remote beaches on Barbados, Guadeloupe and Antigua.

br.

br.

Brown Pelican *Pelecanus occidentalis* L 130cm WS 217cm

The largest bird in the region. Identified by its enormous grey-brown bill (almost the length of its entire body) and throat pouch. The plumage is primarily dark greyish-brown above and richer brown below. Plunge-dives into the ocean, opening its gargantuan mouth to engulf prey.

Vocalisations: the adult emits a low, gruff *hrraa hrraa* during display.

Where to see: possible throughout the region but regularly seen off Dominica, Guadeloupe and Martinique.

Magnificent Frigatebird *Fregata magnificens* L 99cm WS 230cm

Large, black seabird, with long, thin, sharply bent wings, a deeply forked scissor-like tail and a long, thin grey bill that is hooked at the tip. Males have a red gular sac and females a white breast. Immatures have a completely white hood and show a white breast and belly.

Vocalisations: little known but largely silent.

Where to see: coastal and offshore throughout. Codrington Lagoon National Park on Barbuda provides spectacular views of nesting birds.

♂

♀

juv.

Red-footed Booby

Sula sula L 71cm WS 138cm

Oceanic birds best identified by their red feet and large powder-blue bills. Can occur as white (with all-white plumage, black at the wing-tips and along the wing edges) or dark morphs (darker brown upperparts and wings, with a paler brown head and underparts). Juveniles are brownish overall and show a greyish-beige bill and dull yellow feet.

Vocalisations: little known but utters occasional squawking sounds.

Where to see: scarcer than the larger Brown Booby, and often seen from shore flying low.

Brown Booby *Sula leucogaster* L 73cm WS 142cm

Large seabird with long pointed wings and tail. The adults are a deep, dark brown, with a sharply demarcated white belly and vent. In flight, the underwing shows white thickly trimmed with brown. The large, dagger-like bill is yellow, and the feet are yellowish-green. Juveniles are brown with paler underparts and underwings, and dull grey bills.

Vocalisations: little known and largely silent, but squawks, grunts and quacks have been described.

Where to see: the commonest booby in the region, often seen from shore skimming low over the water. Regular off the west coasts of St Vincent, Montserrat and Anguilla.

Masked Booby *Sula dactylatra* L 86cm WS 161cm

Notably larger and rarer than the other two species of booby in the region. White, with a black mask and tail, and black-edged wings. Both the long, chunky bill and large webbed feet are yellow. The juvenile has a brown head, neck, wings and upperparts (crucially, with a noticeable white collar) and largely white underparts.

Vocalisations: silent away from breeding colonies.

Where to see: possible throughout the region but most likely in St Vincent and the Grenadines.

♂

♀

Anhinga *Anhinga anhinga*
L 86cm WS 120cm

A cormorant-like bird with a very long neck and tail. Males show completely black plumage with elegant white streaking on the back and wings. Females have a fawn-coloured head, neck and upper breast. Both sexes have long, straight, sharply pointed orange bills and cream-tipped tails. Commonly known as 'Snakebird', as it swims with its entire body submerged, and only the neck and head visible.

Vocalisations: usually silent in the region.

Where to see: a very rare visitor to the region seen in mangrove wetlands.

American Oystercatcher *Haematopus palliatus* L 42cm WS 76cm

Large, boldly plumaged wader, with a black hood, dark brownish-black upperparts and white underparts. The long, straight, bright red bill and bright orange eye stand out (at close range, a red eye-ring is also visible). The legs and feet are pale pink.

Vocalisations: a repetitive, whistling *pheep-pheep-ph-pheep*.

Where to see: rare on the coast throughout, but more frequent in the north. Shallow tidal pools and wet grasslands.

Black-necked Stilt *Himantopus mexicanus* L 38cm WS 75cm

Tall, slender, 'neat' wader. Its snow-white underparts contrast with a jet-black crown, nape, upper back and wings, creating a unique pied appearance. The black face shows a thick white streak above the eye and a white forehead. The long, black bill is extremely thin and slightly upturned. The pencil-thin legs are bright pink and very long.

juv.

Vocalisations: a repeated *wek-wek-wek-wek …*

Where to see: a breeding resident more likely on islands from Guadeloupe north. Prefers fresh and brackish wetlands.

Grey Plover *Pluvialis squatarola*
L 29cm WS 77cm

Similar but bulkier and taller than American Golden Plover (below). Breeding birds show a white crown, black face and underparts, white vent and heavily dappled black-and-silver upperparts. Non-breeders are greyish-brown above and white below with a 'dirty' breast, plus dark crown, pale eyebrow and dark ear-coverts. In all plumages, contrasting diagnostic black 'armpits' can be seen well.

Vocalisations: a three-note *phreeEE-oo-EEEE*.

Where to see: a common migrant, seen on shallow wetlands and coasts.

br.

non-br.

American Golden Plover *Pluvialis dominica* L 26cm WS 69cm

Similar but smaller and slenderer than Grey Plover. Breeding birds show blotched gold and silver over a black crown and upperparts. The underparts and face are black. As in Grey Plover, note the thick white band extending down the neck-sides to the breast and flanks. Non-breeders have similar plumage to juveniles, with pale brown upperparts and dappled off-white underparts, along with a shaded crown and dark ear-coverts. The back and wing feathers are flecked cream. The prominent off-white eyebrow and large black eye remain.

br.–non-br.

br.

Vocalisations: a short, two-note whistle *phlu-weee*.

Where to see: a regular migrant that favours short wet grass such as golf courses.

Semipalmated Plover

Charadrius semipalmatus
L 18cm WS 47cm

A small wader with relatively long, slender wings, a rounded head, masked face, and a short brown bill with an orange base. The crown and upperparts are soft brown, the underparts white. All adult plumages show an obvious white collar, throat and forehead, a yellowish-orange eye-ring and a breast-band (black in breeders, brown in non-breeders). Legs are yellow-orange.

Vocalisations: a short, rising, whistled *ztuh-WEeet.*

Where to see: a very common migrant to coasts, wetlands and mudflats.

br.

non-br.

br.

non-br.

Wilson's Plover

Charadrius wilsonia L 20cm WS 36cm

Resembles non-breeding Semipalmated Plover with its light brown crown and upperparts, white underparts, white eyebrow, collar, throat and forehead, and its brown breast-band (which is black in breeding males). However, Wilson's is noticeably larger with a flatter head, a heavier, longer, all-black bill and pale pink legs. Non-breeders may show an incomplete breast-band.

Vocalisations: a single whistled sharp *thweet.*

Where to see: an uncommon migrant throughout which breeds on islands from Guadeloupe northwards (and on Grenada). Prefers sandy beaches, rocky coastlines and mudflats.

Killdeer *Charadrius vociferous* L 24cm WS 61cm

Similar in plumage colour and general pattern of markings to the more common Semipalmated Plover, but a significantly taller bird with longer, paler legs, a longer tail and a black double breast-band. There is a thin, all-black bill and chestnut rump (visible in flight).

Vocalisations: a repeated, high-pitched, squeaky *twiddleyeee.*

Where to see: an uncommon migrant to the more northern islands. Prefers manicured lawns and recently ploughed agricultural fields.

br.

non-br.

Snowy Plover *Charadrius nivosus* L 16cm WS 43cm

A small, pale plover with an incomplete breast-band that never meets in the centre. It has black ear-coverts and forecrown (brown in females and juveniles). The only small plover in the region with dark grey legs.

Vocalisations: a crisp *prrrp.*

Where to see: an infrequent migrant throughout. Also, a breeding resident on some of the northern islands (particularly Anguilla and Guadeloupe). Prefers sandy beaches and mudflats.

Collared Plover

Charadrius collaris L 15cm WS 37cm

A small, delicate plover with a chestnut-tinged crown, head and upperparts, along with a thin black breast-band. Pale yellowish legs, a black frontal bar, partial eye-stripe, dark ear-coverts and a very thin black bill also distinguish this species. Juveniles lack a breast-band.

Vocalisations: a continuous garbled melee of rapid *prrp* notes, whistles and trills.

Where to see: rare, with just a few individuals seen annually on Barbados and Grenada.

Southern Lapwing *Vanellus chilensis* L 35cm WS 32–38cm

A large, heavy-set, greyish-brown plover with thin, black, lanceolate plumes extending from the base of the crown, a bright red eye and eye-ring and a black forehead and throat. There is a broad black breast-band, white belly and a bronzed purplish-green iridescence on the back and wings. The bill is pink with a black tip, and the legs are red. In flight, the wings are black-and-white.

Vocalisations: a 'yapping' alarm call (like a small dog).

Where to see: has bred on Barbados and Grenada. Prefers open plains and tilled agricultural fields.

Upland Sandpiper *Bartramia longicauda* L 29cm WS 66cm

A large wader with a rather rotund body, lengthy neck and tail, and small head (which shows a faint white eye-ring). The long, slim, slightly drooped bill is dark above and yellow below. The brown upper-parts and wings have a checked appearance due to pale-edged feathers. The crown, head, foreneck, breast and flanks all show brown streaking, while the chin, belly and vent are white and unmarked.

Vocalisations: a rolling *prrrrrrruuhiiiiip*.

Where to see: a regular migrant to Guadeloupe (less so, Barbados).

Prefers open fields and grassland (such as golf courses).

Whimbrel *Numenius phaeopus* L 43cm WS 83cm

A very large wader with a long, downcurved, brown bill (the mandible is often tinged pink). The back and wings show a checked pattern of beige and brown, while the paler face, neck and breast are streaked. The crown is dark with a pale central stripe and the face shows a cream eyebrow and brown eye-stripe. Flanks are barred.

Vocalisations: a loud sharp *wheeeee---peeep-peeep-peeep*.

Where to see: a regular autumn migrant in small numbers. Often forages on exposed reefs and tidal pools and on coastal marshland.

Marbled Godwit *Limosa fedoa* L 45cm WS 75cm

This large, primarily tawny-brown wader has a distinctive, long, upturned, two-toned bill (which is pinkish-orange at the base with a black tip) and a pale-cream eyebrow. The upperparts are darker and speckled with chestnut, while the underparts are pale and heavily barred with reddish-brown. Distinguished from Hudsonian Godwit (below) by its larger size and absence of white on the rump and tail.

Vocalisations: regular, repetitive, up-slurred calls *arr-ehh arr-ehh.*

Where to see: a very rare visitor to the region at beaches and tidal pools.

Hudsonian Godwit *Limosa haemastica* L 39cm WS 73cm

A large wader with a long, upturned, two-toned bill (pinkish-orange from the base with a black tip) and prominent pale eyebrow. Non-breeders are greyish-beige (darker above). Breeding birds show a finely streaked pale face and head, dark crown, mottled dark brown upperparts and barred rufous-orange underparts. In flight note the broad black terminal band on an otherwise white tail and the white bar on the wing.

Vocalisations: a short, distinctive *yap.*

Where to see: a far more common migrant than Marbled Godwit. Most likely tidal pools, lagoons and wet grasslands on Martinique, Guadeloupe and Barbados.

Ruddy Turnstone *Arenaria interpres* L 23cm WS 54cm

This short, stocky, medium-sized wader is particularly striking in breeding plumage, when it shows a bright chestnut-orange and black montage of colour splayed across the back and wings, and an elaborate black-and-white breast and facial pattern. Non-breeders are much duller, with the orange and black being largely replaced by grey and brown. Belly and vent are white across all plumages.

Vocalisations: a trio of staccato *thoops*.

Where to see: a common migrant wader often seen perched on jetties and piers, as well as on tidal pools and seaweed-strewn beaches.

br.

non-br.

Red Knot *Calidris canutus* L 24cm WS 50cm

This medium-sized, stocky and chunky wader has grey upperparts and white underparts in winter, with faint grey streaking on breast and flanks. There is a broad white eyebrow, longish black bill and greenish-yellow legs. In flight, the pale rump is finely barred with white and grey. Breeding birds have rufous faces and underparts, and an elaborate mottled pattern of rufous, chestnut, grey and black on the crown and upperparts, and black legs.

Vocalisations: a refined and double-noted *chuuhp-chuuhp*.

Where to see: this uncommon migrant is most likely to be seen on Barbados in September on coastal wetlands and mudflats.

Ruff *Calidris pugnax* L 24cm WS 52cm

Males are roughly the size of Greater Yellowlegs, with females the size of Lesser Yellowlegs. The plumage of non-breeders is similar to both, but the upperparts of Ruff show clearly delineated large feathers, and the legs are orange not yellow. White feathering shows at bill base, and the rest of the underparts smudged with grey-brown. Breeding male Ruff is unmistakable, with an ostentatious orange, black, white or purple mane starting at the crown and extending to the back and breast. In flight, a white 'V' is obvious on the rump.

Vocalisations: a seldom-uttered *gnyahh gnyuhp*.

Where to see: a rare Old World vagrant to wetlands on Barbados and Guadeloupe.

Stilt Sandpiper *Calidris himantopus* L 22cm WS 43cm

Non-breeders resemble Lesser Yellowlegs due to the greyish-brown upperparts, white underparts and notable streaking on the pale head, neck and breast. Stilt Sandpiper is slightly smaller with a lengthy black bill (drooped at the tip), long, broad, white eyebrow and greenish-yellow legs. Breeding birds have checked rufous-brown upperparts, heavily barred underparts and a rusty-orange crown and ear-coverts.

Vocalisations: a hoarse *kwerrp.*

Where to see: this common migrant is often seen in small flocks on wetlands and mudflats alongside yellowlegs and dowitchers.

br.

(L) non-br.

Sanderling *Calidris alba*

L 20cm WS 37cm

At a distance this chunky but small wader can appear totally white in non-breeding plumage. The face and underparts are white, but the upperparts are a very pale silverish-grey with subtly dark feathering in the wing. Breeders show spangled rufous and black on their head, upperparts and breast. Juveniles have a buff-tinged crown, head and breast, along with black and greyish-white checkered upperparts.

Vocalisations: a short single *pip* (occasionally *pih-pip*).

Where to see: this common migrant has a fondness for sandy beaches and is often seen scurrying at the water's edge.

Dunlin *Calidris alpina* L 19cm WS 37cm

Among the largest of the small sandpipers, Dunlin has a long, drooping, black bill (most pronounced at the tip), black legs and a hunched appearance. Non-breeders show a faint white eyebrow and are a plain greyish-brown above and white below, with grey shading on breast and flanks. Breeding birds show heavy rufous on the crown and back, an obvious black belly patch and a finely streaked pale head and breast.

Vocalisations: a thin *rhleeep.*

Where to see: an uncommon migrant usually seen among other species on shallow wetlands or mudflats.

br.

non-br.

Baird's Sandpiper

Calidris bairdii L 16cm WS 38cm

Resembles Least Sandpiper due to its brown head and upperparts and small size (most other smaller sandpipers are a plain brownish-grey). The breast has brown streaks. Baird's is slightly larger than Least and has wings that project beyond tail at rest, and black legs. Breeding birds show a significant amount of black in the otherwise tawny-brown upperparts and crown.

Vocalisations: a single flat *preep*.

Where to see: a rare visitor to Barbados and Guadeloupe. Prefers feeding in inundated short grass.

Least Sandpiper *Calidris minutilla* L 14cm WS 34cm

The smallest sandpiper (or 'peep') in the region. It is notably browner than other peeps on the head and upperparts, with brown streaking on the breast and yellow legs. In breeding condition, the plumage becomes rufous-tinged, but the legs remain yellow. The bill is dark brown and slightly decurved.

Vocalisations: a high-pitched *chreee*.

Where to see: a common migrant to wetlands and exposed mudflats (sometimes beaches).

White-rumped Sandpiper *Calidris fuscicollis* L 17cm WS 37cm

Another small sandpiper, in which non-breeders show a white eyebrow and are largely dull grey above and white below, with light streaking on the breast. White-rumped is larger than some, however, and the wing-tips project significantly beyond the tail at rest. In flight, an all-white rump patch is obvious. Breeding birds are mottled rufous, chestnut and black on the back and wings, and show a rufous-tinged crown and ear-coverts. The bill typically shows a pale orange base.

Vocalisations: a high-pitched *tseep-tseep*.

Where to see: a common migrant to mudflats, shallow ponds and wetlands.

br.–non-br.

Buff-breasted Sandpiper *Calidris subruficollis* L 19cm WS 45cm

Similar in plumage and colour to the regionally more common Pectoral Sandpiper, but slightly smaller and slimmer, with unstreaked buff underparts (paler on the vent). The species has a short, sharp, all-black bill, a typically erect posture, and a small round head, making it more reminiscent of a plover than a sandpiper.

Vocalisations: a single *tseep* in flight.

Where to see: an uncommon but reliable migrant to Barbados, Guadeloupe and Martinique. Favours grassland, fields, pastures and golf courses.

Pectoral Sandpiper *Calidris melanotos* L 21cm WS 41cm

A medium-sized, chunky wader with a dark crown, along with dense brown streaking on an otherwise white head, neck and breast. Note the sharp and obvious demarcation where the streaked breast meets the white belly; the vent is also white. The back and wings are brown with pale and chestnut-edged individual feathers. In all plumages, has a white eyebrow, a two-toned bill and greenish-yellow legs.

Vocalisation: a short, clear, chirped *prreet*.

Where to see: a common migrant to mudflats and the banks of shallow freshwater and brackish wetlands and inundated turf. Can be rather approachable.

juv.

br.

Semipalmated Sandpiper *Calidris pusilla* L 15cm WS 35cm

Non-breeding birds are greyish-brown above and white below with dense greyish-brown streaking on the breast flanks. The grey head shows a white eyebrow. Breeders have dark streaking to the breast and a subtle rufous tinge to the upperparts and head. The straight, stout, black bill and black legs are consistent across all plumages. At very close range, the toes show partial webbing.

Vocalisations: a short, sharp *cherrt.*

Where to see: an extremely common migrant to the region. Regularly seen in large flocks on coasts and exposed mudflats.

non-br.

br.

Western Sandpiper *Calidris mauri* L 16cm WS 36cm

Non-breeders are almost identical to Semipalmated Sandpiper, though Western has paler grey upperparts and a finely streaked white breast. Western is longer legged, with a longer black bill, typically drooped at the tip. Breeders show much more rufous to the upperparts and head than Semipalmated, with heavy chevrons on the flanks.

Vocalisations: high-pitched 'cheeping' whistles and *preet* notes.

Where to see: not nearly as numerous in the region as Semipalmated, but a reliable migrant to wetlands, mudflats and coasts.

non-br.

br.

Short-billed Dowitcher *Limnodromus griseus* L 27cm WS 48cm

A medium-sized wader with an impressive bill. The head and upperparts are greyish-brown with white underparts showing a grey-shaded breast and dark barring on the flanks, vent and undertail. In flight, heavy barring also shows on the otherwise white lower back and rump. Breeding birds have mottled black-and-chestnut upperparts and buffish-orange underparts. A white eyebrow, dark eye-stripe and yellow-green legs are present across all plumages. The long bill probes into the mud with a rapid, sewing machine-like motion.

Vocalisations: a rapid, 'laser gun-like' *tchuu-tchuu-tchuu.*

Where to see: a regular migrant typically seen on coasts and mudflats.

br.–non-br.

non-br.

Long-billed Dowitcher *Limnodromus scolopaceus* L 28cm WS 49cm

Almost identical to Short-billed Dowitcher, but longer legged with generally darker plumage and a longer, green-based bill. At rest, the wing-tips do not extend beyond the tail (those of Short-billed typically do). The species are best distinguished by voice.

Vocalisations: single-noted and slower than Short-billed – *cheep … cheep … cheep*

Where to see: a rare migrant to the region.

Wilson's Snipe *Gallinago delicata* L 27cm WS 44cm

A chunky, medium-sized wader with a long, straight, dark bill and short greenish-yellow legs. The intricate head and face pattern includes a striped crown, and tawny-coloured cheeks with two pale, horizontal bands framing the eye. The upperparts are mottled brown, beige and black, and show pale streaking. The breast and flanks are heavily barred, contrasting with the unmarked creamy-white belly and vent.

Vocalisations: a harsh, husky rasping croak on flushing, when it careers away rapidly and in zigzags.

Where to see: a common migrant to inundated grassland near water.

Wilson's Phalarope *Phalaropus tricolor* L 23cm WS 38cm

Non-breeders are plain, though scaly, grey above and white below, with a white eyebrow and forehead, a long, very thin, pointed, black bill, and yellow legs. Breeding birds show black legs, broad, rufous streaks running along bluish-grey upperparts (duller in males), and chestnut neck-sides and foreneck. The remaining underparts are white. Females show a thick, black eye-stripe extending round to the base of the head and onto the neck (this is brown and less obvious in males). Swims and spins in tight circles when feeding in water.

non-br.

Vocalisations: a repeated *phweh-yahweh-phweh*.

Where to see: a rare visitor to salt ponds on Anguilla, Guadeloupe and Barbados.

br.

juv.

Spotted Sandpiper

Actitis macularius L 19cm WS 38cm

The only sandpiper with large, dark brown spots on white underparts. The upperparts and head are finely barred pale brown. The head also has a white eyebrow and white, incomplete eye-ring. Non-breeders and juveniles lack spots and barring is much reduced on the head and upperparts. A vertical white, finger-like patch is present at the bend of the folded wing. Regularly bobs its tail and rear end. Very short, rapid but stiff wingbeats in flight.

Vocalisations: a repeated, high-pitched, ascending *peep* increasing in rapidity. Sometimes a rolling *preeeeet*.

Where to see: a common migrant to almost all bodies of water.

Solitary Sandpiper *Tringa solitaria* L 20cm WS 57cm

The similar size, largely brown upperparts, white underparts and tendency to bob its tail, are very similar to non-breeding Spotted Sandpiper. Solitary is slightly larger and longer, with a darker brown back and wings, delicately sprinkled with small white flecks, an obvious white eye-ring, and greyish-green legs. The white head and breast are lightly shaded with brown, replaced by heavy brown streaking in breeders. The belly and vent are completely white. In flight, dark brown and white barring shows on the tail.

non-br.

Vocalisations: *phwee-wee-weep.*

Where to see: a regular migrant to shallow freshwater or brackish ponds, mangrove swamps and other wetlands.

br.

Willet *Tringa semipalmata* L 37cm WS 61cm

A large, heavy-set, rather plain wader with a greyish-brown head and upperparts and greyish-white underparts. The long, thick, straight, black bill is paler at the base and the long legs are greenish-grey. Breeders show darker, brown-speckled upperparts, a streaked head and neck, and a heavily barred breast. In flight, note the 'zebra pattern' of the wings created by a broad white wing-bar between two shorter black bars.

Vocalisations: a repeated *tree-leep tree-leep*.

Where to see: a regular migrant to beaches, rocky shorelines, freshwater marshes, mudflats and mangroves.

non-br.

br.

Lesser Yellowlegs *Tringa flavipes* L 24cm WS 62cm

Essentially a smaller, slimmer version of the Greater Yellowlegs (below) with almost identical plumage and leg colour. Lesser has a shorter straight black bill, roughly the length of its head.

Vocalisations: a single (occasionally double) *tu*.

Where to see: a very common migrant (far commoner than Greater) favouring expanses of mudflats, freshwater ponds and other shallow wetlands.

Greater Yellowlegs *Tringa melanoleuca* L 31cm WS 72cm

A large wader with greyish-brown upperparts and wings heavily flecked with white, greyish-brown streaking on its white head, neck and breast, an unmarked belly and vent, and long, bright-yellow legs. The slightly upturned bill is pale greenish-grey at the base and dark brown at the tip. Breeding individuals have denser, darker streaking on the head and neck, black flecks on the breast and upper belly, black splotches on the upperparts and an all-black bill.

Vocalisations: typically, three to four *tu-tu-tu* notes.

Where to see: a regular migrant to freshwater ponds, marshes with shallow open water and mangroves.

Brown Noddy *Anous stolidus* L 38–45cm WS 75–86cm

A sleek tern with warm chocolate-brown plumage, long pointed wings and a slender tail. Note the thin white crescent below the eye, pale greyish-white cap (absent in juveniles) and the relatively long, thin, black bill. Erratic flight at varying heights over the ocean. Regularly perches on buoys and boats offshore.

Vocalisations: a crow-like *caw*.

Where to see: oceanic waters throughout the region, especially off western Grenada, St Vincent, Anguilla and Saint Martin.

Black-headed Gull *Chroicocephalus ridibundus* L 41cm WS 101cm

Very similar to Laughing Gull but with shorter wings showing more white in the tips, significantly paler grey upperparts and a thinner maroon-coloured bill. Breeding birds have a partial brown hood (with a white nape), while non-breeders have white heads with an obvious dark spot behind the eye. The legs are red.

Vocalisations: a frequent, screaming, down-slurred *creee-arrr*.

Where to see: a very rare visitor to Barbados, Guadeloupe, St Lucia or Grenada.

non-br.

br.

Laughing Gull *Leucophaeus atricilla* L 41cm WS 109cm

Breeding birds have full black hoods with a split white eye-ring, red bills and legs. The neck and underparts are snow-white, and the back and wings a rich, dark grey (with heavy black at the wing-tips). Non-breeders are paler, the black hood being replaced by grey shading on the nape and rear of the crown. The bill and legs are black. Juveniles have brownish-grey plumage with dark bills and legs.

Vocalisations: an often-repeated set of loud *kek-kek-kek* and *OO–ah* sounds.

Where to see: common throughout.

br.

br.–non-br.

juvs.

Ring-billed Gull *Larus delawarensis* L 48cm WS 125cm

Breeding birds have a white head, neck, underparts and tail, contrasting with pale grey upperparts. Non-breeders are similar, with brown streaking on the head and neck. The most consistent features are a black ring encircling the thick yellow bill towards the tip and bright yellow legs.

Vocalisations: loud, repetitive *uk-uk-uk* calls and longer screams.

Where to see: uncommon throughout, but most frequent off Anguilla, Barbados and Guadeloupe.

non-br.

1st win.

br.

Lesser Black-backed Gull

Larus fuscus L 56cm WS 138cm

A large gull, which in breeding plumage has a completely white head, neck, underparts and tail, and a dark grey back and wings – which are long, pointed and tipped black. Non-breeders show a soft brown streaking on the head and neck. The legs and chunky bill are yellow, the latter showing a red dot towards the tip.

Vocalisations: similar but deeper than Herring (below).

Where to see: uncommon visitor to coasts throughout.

juv.

non-br.

Herring Gull *Larus argentatus* L 59cm WS 135cm

Similar in size to Lesser Black-backed, with almost identical plumage and markings in both breeders and non-breeders. However, Herring has a much paler grey back and wings, and pale pink legs.

Vocalisations: a subtly descending, raucous, laughing *arr-arr-arr-arrr*.

2nd win.

br.

Where to see: irregular but possible throughout the region.

Sooty Tern *Onychoprion fuscatus* L 40cm WS 88cm

Any tern in the region with very dark upperparts and white underparts is likely to be a Sooty or Bridled Tern (next page). Both show a glossy black cap and nape, long elegantly pointed wings, a deeply forked tail and a black bill and legs. Sooty is larger and blacker, with the white forehead extending up from the base of the bill diagnostically stopping at the eye. Juveniles are dark brown with white flecks.

Vocalisations: typically, *wideawake-wideawake* and various *yip* calls, given in flight.

Where to see: possible at sea throughout. Breeds on Anguilla and surrounding cays, as well as uninhabited islands in the Grenadines.

Bridled Tern *Onychoprion anaethetus* L 31cm WS 79cm

Almost identical to Sooty, but smaller, with duller greyish-black upperparts, tinged with brown. The white forehead extends from the base of the bill to well behind the eye, like a long white eyebrow. Non-breeders show more white in the crown than non-breeding Sooty. Juveniles are grey-brown above with pale barring and a streaked crown.

Vocalisations: a dog-like *ark* plus 'squeaky-toy-like' sounds.

Where to see: more commonly sighted than Sooty, often perching on buoys and piers. Breeds on Anguilla and surrounding cays, as well as in the Grenadines.

br. / non-br.

br.

Least Tern *Sternula antillarum* L 22cm WS 50cm

Adults are very small, with otherwise typical tern plumage. Note the bright yellow bill, tipped with black, the obvious visor-shaped white patch above the bill, and the yellow legs. Non-breeders show white streaking to the black forecrown and a yellowish-black to black bill, but the white 'visor' remains visible. Juveniles show brownish dappled wings, back and crown.

Vocalisations: a high-pitched, two-note *ki dik* and a *zrceep* alarm call.

Where to see: common on undisturbed beaches and open mudflats in St Kitts and Nevis, St Maarten, Barbuda and Montserrat.

Gull-billed Tern *Gelochelidon nilotica* L 35cm WS 100cm

A medium-sized tern with a relatively short, thick, black bill and long black legs. Breeders show a black cap and nape, pale grey upperparts, white underparts and a short, slightly forked tail. Non-breeders have a pure white head and black ear-coverts. Juveniles are similar but show sparse, faint, brown streaking on the crown, back and wings.

Vocalisations: a two-note *kee-reek*.

Where to see: possible throughout, preferring marshes, freshwater ponds and mudflats.

non-br.

br.

Caspian Tern *Hydroprogne caspia* L 52cm WS 134cm

The largest tern in the region, and told from the similar, much more common Royal Tern by its heavier build, much larger, thicker and richer red bill, and longer wingspan (crucially showing a black wedge on the underside of the wing-tips). Breeding birds show an all-black cap with shaggy black crest (becoming interspersed with grey in non-breeders). Note also the very pale grey back and wings, white underparts, black legs and slightly forked tail. Upperparts on juveniles are punctuated with sporadic dark chevrons.

Vocalisations: loud *rah* and *rau* sounds.

Where to see: possible throughout, but more common in the north.

Roseate Tern *Sterna dougallii* L 38cm WS 76cm

A medium-sized tern with very pale grey upperparts (appearing white at a distance) and all-white underparts. Breeders show a black cap and nape (with the black only behind the eyes in non-breeders) and thin black on the forewing; underparts sport a very pale pink flush. The tail is forked with all-white tail-streamers extending well beyond the wing-tips when the bird is at rest. The bill is black to reddish-black with a heavy orange base in breeders, while the legs are orange-red. Juveniles show dark-edged feathering across the upperparts and black legs.

Vocalisations: high-pitched *pink* and *ki-rik* calls.

Where to see: found along southern and western coastlines in summer and breeds on St Vincent.

Black Tern *Chlidonias niger* L 26cm WS 61cm

A small, dark tern, which in breeding plumage shows a glossy black hood and underparts and largely sooty-grey upperparts. Non-breeders have a white forehead, nape and underparts, with pale grey on the upper flanks. Upperparts are medium grey. The crown is dark and there is a dark ear patch. This tern shows a twitchy and jerky flight style and plucks prey delicately from the water surface.

Vocalisations: silent away from breeding colonies.

Where to see: very rare. Forages in brackish mangrove swamps and at sea.

Common Tern *Sterna hirundo* L 36cm WS 77cm

The underparts are white, the upperparts and wings medium grey with dark shading at the wing-tips. The tail is forked and shows dark-edged white tail-streamers that do not extend beyond the wings at rest. Breeders have bright orange bills, a black cap and nape, and orange legs. In non-breeders, the bill is black, the cap is restricted to behind the eyes, and the legs are a dull reddish-black.

Vocalisations: a sharp, variable, frequently uttered *kip* and *keee-er-ree*.

Where to see: most likely in summer and winter along southern and western coasts.

Arctic Tern *Sterna paradisaea* L 35cm WS 81cm

Broadly resembles both Common and Roseate in all plumages, but slimmer, shorter legged and has longer tail-streamers than Common with darker grey upperparts than Roseate. Arctic also has a more rounded head and shorter neck. Breeders show a bright red bill and legs.

Vocalisations: very similar to, but higher pitched than, Common.

Where to see: a very rare visitor most likely on Guadeloupe.

br.

br.

Sandwich Tern *Thalasseus sandvicensis* L 41cm WS 96cm

A medium-large tern – but smaller than Royal – with white underparts, a pale grey back and wings, and a black cap with a shaggy black crest (in non-breeders, the cap is reduced, starting behind the eye). The long, thin bill is black with a yellow tip, and Sandwich is the only tern in the region with this feature. The tail is forked and the legs black.

Vocalisations: a loud, harsh *keee-rrikk*.

Where to see: uncommon. Most often seen inshore.

br.

non-br.

Royal Tern *Thalasseus maximus* L 48cm WS 117cm

This second-largest tern in the region has a large, thick, bright orange bill, a full black cap and shaggy black crest (much reduced in non-breeders, starting behind the eye). The back and wings are pale grey, the wing-tips dark, with white underparts, black legs and a shallow forked tail. Juveniles have dappled light brown upperparts, a yellow bill and yellowish legs.

Vocalisations: a loud *keet-keet* or *kleer-kleer*.

Where to see: common close to shore.

br.

non-br.

(far L) juv.

Long-tailed Skua (Long-tailed Jaeger) *Stercorarius longicaudus*

L 51cm including tail-streamers WS 111cm

A small jaeger the size of a Laughing Gull with similar plumage to Pomarine and Parasitic (although, crucially, no pale patch visible on the underwing). Delicate and streamlined with a tern-like flight on shorter wings. Breeding birds have two very long, thin feathers extending beyond the tail. The bill is short and slim. Mostly silent away from breeding sites.

Where to see: a pelagic species most reliable off Guadeloupe and Dominica in summer.

Pomarine Skua (Pomarine Jaeger)

Stercorarius pomarinus L 49cm including tail-streamers WS 131cm

The most powerfully built jaeger, with a thick pink-and-black bill. Dark morphs have sooty grey-black plumage while pale morphs have a dark brown head, soft yellow neck, sooty-grey upperparts and largely pale underparts (often with a dark breast-band and barred flanks). Breeding birds have extended spoon-like central tail feathers. Juveniles are brown with significant barring.

Vocalisations: feeding birds utter *wich-yew … wich-yew* and *week-week-week* calls.

Where to see: the most regular jaeger in the Lesser Antilles, particularly off the coasts of Guadeloupe and Dominica from October–April.

Arctic Skua (Parasitic Jaeger) *Stercorarius parasiticus*

L 44cm including tail-streamers WS 118cm

Both morphs are very similar to Pomarine Jaeger but Parasitic is slimmer-bodied, with a smaller head and thinner bill and more rapid wingbeats. Breeding birds have long, slim, central tail feathers protruding beyond the tail and culminating in a sharp point.

Vocalisations: a harsh *kyew kyew* flight call.

Where to see: rare throughout, but relatively frequent off Guadeloupe from March to June.

South Polar Skua *Stercorarius maccormicki* L 53cm WS 135cm

A large, heavy-set barrel-chested, gull-like seabird, with a thick, black bill and dark brown plumage. In flight, appears hunched with rounded wings. There is an obvious white patch towards the wing-tip and a short, square tail.

Vocalisations: usually silent away from its breeding grounds.

Where to see: possible throughout, especially during spring, particularly over deep ocean waters off Guadeloupe and Dominica.

Ashy-faced Owl *Tyto insularis* L 30cm WS 24-28cm

Very similar to the well-known and cosmopolitan Barn Owl, but much darker. A dark greyish-brown crown is paired with dark brown upperparts and wings that are heavily interspersed with tawny and black feathering (occasionally flecked white). The underparts are warm chestnut-beige, and the broad facial disc is a soft greyish-maroon. Nocturnal.

Vocalisations: a high-pitched, hair-raising shriek.

Where to see: the only resident owl in the region. Found on Dominica, St Vincent and the Grenadines and Grenada. Frequents residential areas and open woodland.

Osprey *Pandion haliaetus* L 57cm WS 152cm

This largest raptor of the region has dark chocolate-brown upperparts and white underparts. The head is white with a thick, dark-brown eye-stripe (females also have a brown shaded crown and breast-band). The white undertail and underwings are barred with pale brown, the wings also having a dark brown patch at the bend.

Vocalisations: a very high-pitched *kyeee … kyeee.*

Where to see: regular migrant over or near large waterbodies.

Hook-billed Kite *Chondrohierax uncinatus* L 41cm WS 80cm

A medium-sized raptor that resembles a flying crucifix thanks to its broad wings, long tail, and outstretched neck and head. Males have a slate-blue head and upperparts with greyish-blue barring on the white breast, belly and underwing. Females have the slate-blue head, but the upperparts are rufous-brown, and the underparts and underwings are heavily barred with chestnut. Juveniles show white cheeks, brown upperparts, and white with brown barring below.

Vocalisations: a double-noted whistle.

Where to see: regularly seen soaring above the dry forests of Mt Hartman National Park, Grenada.

juv.

♀

Common Black Hawk *Buteogallus anthracinus* L 53cm WS 117cm

A large, all-black raptor with a big, bright yellow bill (tipped with black) and bright yellow legs. In flight, shows a prominent thick white band on the tail, the tip of which is also edged with white. Juveniles show heavily streaked, pale underparts. Often perches near forest streams, preying on crabs.

Vocalisations: multi-note whistles – *phee-pheEE-PHEEE-PH-phe-phe-phe*.

Where to see: soars over dense forests on St Vincent (especially La Soufrière) and Grenada.

Broad-winged Hawk *Buteo platypterus* L 37cm WS 85cm

A medium-sized hawk with broad, rounded wings and a heavily banded tail. The upperparts and head are typically dark to medium brown, the underparts pale with brown barring and splotching on the breast, belly and darkly edged underwings. There is notable variation in size and plumage among the resident subspecies: e.g., subspecies *insulicola* (Antigua) is smaller and paler than subspecies *rivierei* (St Lucia, Martinique and Dominica).

Vocalisations: a lengthy high-pitched *tkeeeeee.*

Where to see: a migrant throughout, as well as a common breeding resident on St Lucia, Martinique, Dominica, Grenada, Antigua and St Vincent.

Frequents dense forests, dry scrubland and even city centres.

Red-tailed Hawk *Buteo jamaicensis* L 56cm WS 127cm

A large raptor showing long, broad, rounded wings and a large, fanned tail in flight. There is much variation in the plumage, but the majority have a dark head and crown, mottled dark brown and rufous-chestnut upperparts, and pale underparts punctuated by a dark-streaked band across the belly. Upperside of tail shows reddish-brown. Juveniles are grey and white.

Vocalisations: a coarse descending *shEEEEeeer.*

Where to see: some northern islands, where commonest on St Kitts and Nevis in most habitats.

juv.

Ringed Kingfisher *Megaceryle torquata* L 40cm WS 68cm

A very large, crow-sized kingfisher with a blue crest, head and upperparts, and an obvious thick white collar. Males have entirely rufous underparts (save for a white vent), while females have a blue breast, thin white breast-band and a rufous belly and vent.

Vocalisations: a long, sustained rattle, higher than Belted (next page).

Where to see: Dominica and Guadeloupe, where it is often seen perched on waterside trees and telephone wires.

♀

♂

Belted Kingfisher *Megaceryle alcyon* L 31cm WS 53cm

A medium-sized blue-and-white kingfisher with a shaggy blue crest. Males have a blue head and upperparts and white underparts broken by a single blue breast-band. Females are similar but with rust-coloured flanks and a broad, rusty belly-band. Both sexes have a thick, white collar.

Vocalisations: a machine gun-like rattle.

Where to see: a regular migrant to coast and wetlands.

Channel-billed Toucan *Ramphastos vitellinus* L 51cm WS 60cm

Large and colourful with an enormous, slightly decurved, jet-black bill. The crown, upperparts, breast and belly are black, with a rufous rump and there is a rufous breast-band and vent. The white foreneck has a large, yellow patch. There is pale blue bare skin around the eye and pale blue shows around the base of the bill.

Vocalisations: a buzzy, high-pitched *keeeuhhhk*.

Where to see: introduced to Grenada, where it can be seen in the Grand Etang Forest Reserve.

Guadeloupe Woodpecker *Melanerpes herminieri* L 27cm WS 42cm

A medium-sized, black woodpecker with a maroon-tinged throat, breast and belly. The straight, stout bill and legs are black. Juveniles are brownish-black with orange underparts.

Vocalisations: a barked *KyAark*.

Where to see: endemic to Guadeloupe, where it is found in all forest types. Common in the Basse Terre region, with regular sightings in picnic areas such as Cascade aux Ecrevisses.

American Kestrel *Falco sparverius* L 26cm WS 56cm

This small falcon has rufous-chestnut upperparts, with grey on the crown, and paler underparts. The entire body is spotted and blotched with black. The pale face shows two distinct, thick, black, vertical lines (one running through the eye). The long rufous tail is multi-banded in females, with a single, black subterminal band in males. Sexes also differ with grey in the wings of males absent in females.

Vocalisations: a series of *kieh-kieh-kieh-kieh-kieh* notes.

Where to see: often seen hovering over roadsides and grass verges or perched on telephone wires.

Merlin *Falco columbarius* L 28cm WS 63cm

Similar to Peregrine Falcon, but much smaller and less robust, with thinner, less obvious 'sideburns'. Both sexes have dark crowns, with females showing brown upperparts and males bluish-grey. The cream underparts are heavily streaked, and the underwings heavily barred. A dark tail shows three to four prominent pale grey bands.

Vocalisations: a rapid *tuh-tii-tkitkitkitkitkitkitki.*

Where to see: a regular autumn migrant to wetlands and mudflats.

Peregrine Falcon *Falco peregrinus* L 46cm WS 102cm

A large, powerful, robust falcon with a dark masked face complete with prominent, thick, black 'sideburns'. There is dark, horizontal barring on its pale cream underparts and underwing. The upperparts are slate-blue. Juveniles are brown above with heavily streaked cream underparts.

Vocalisations: a repeated *kyyyyuuuuup--kyyyyyuuuup* given from perched position.

Where to see: a regular migrant that has also nested on Dominica. Often perches high on sea cliffs, surveying for prey.

juv.

Red-necked Parrot *Amazona arausiaca* L 37cm WS 46cm

A chunky, bright green parrot with a bluish face, red crescent-shaped throat patch and a yellowish-green vent. In flight, the upperside of the wing shows red and yellow, with dark blue wing-tips. Underwing is blue. The tail has a broad yellow band. The duller juveniles lack the red throat patch.

Vocalisations: high-pitched trills, squawks and squeals.

Where to see: by far the commoner of the two endemic Dominican parrots. Flocks can be seen regularly along the forested Syndicate Nature Trail, and in the Carholm and Indian River regions.

St Lucia Parrot *Amazona versicolor* L 43cm WS 48–50cm

The only parrot on St Lucia has an overall rich green plumage, highlighted by a blue face and forecrown. The nape, neck-sides and upper back are heavily scaled. A crimson bib bleeds extensively downwards, ensuring the scaled feathering of the breast and belly is a mottled blend of reds and greens. In flight, the wings show brilliant blues and reds, with dark, navy-blue wing-tips. The tail has a broad lemon-yellow band. Juveniles show less blue in the face and have a duller crimson bib.

Vocalisations: a rolling yelp – *preeearrgh … preeearrgh.*

Where to see: endemic to St Lucia, where it is primarily seen in mature forest. The observation area in Des Cartiers Rainforest can produce sightings of birds at eye level.

Imperial Parrot *Amazona imperialis* L 48cm WS 55–60cm

The largest of the world's *Amazona* parrots has dark plumage with a scaly, brownish-purple head, neck, breast and belly, dull green back and wings, and a blue-tipped, dark brown tail. In flight, only the bluish-black wing-tips appear to move, effectively distinguishing Imperial from Red-necked Amazon (in which the entire wing rapidly flaps) even at distance. Juvenile Imperials are similar but show a duller green nape and neck.

Vocalisations: like an old, creaky metallic gate.

Where to see: endemic to Dominica. One of the rarest species on the planet, with the last remaining individuals inhabiting the remote summits of the moist montane forests of Morne Diablotin National Park and Morne Trois Pitons National Park. Seen in its largest numbers on the eastern and north-eastern slopes of Morne Diablotin. Visit in the early morning and listen for the bird's unique call.

Orange-winged Parrot *Amazona amazonica* L 31cm WS 43–46cm

A medium-sized parrot with soft, green plumage and a distinctive yellow forehead, cheeks and throat. In flight, an orange wing patch and yellow-tipped tail are clearly visible.

Vocalisations: when perched, *cluck, screeke* and *scree* calls. In flight, *klyuck-klyuck-klyuck-klyuck.*

Where to see: an introduced species breeding in small numbers in Belleville, Barbados and some forested regions on Grenada. A larger population is more widespread on Martinique in Fort-de-France parks and residential areas.

St Vincent Parrot *Amazona guildingii* L 44cm WS 49–52cm

Majority of plumage is a highly unusual bronzed brown, against which a greyish-white face and crown stand out well. The nape, head-sides and throat are tinged bluish-grey and the large bill is pale. The upperside of the tail is rich blue, with an orange base and a thick, bright yellow terminal band. The undertail is lemon-yellow.

Vocalisations: a scratchy double note followed by a bleating ring.

Where to see: endemic to St Vincent, where it favours mature, moist forests. Frequently seen at dawn and dusk in the Buccament Valley.

Rose-ringed Parakeet

Psittacula krameri L 40cm WS 45cm

The only parakeet in the region has bright, yellowish-green plumage, a long turquoise-blue tail and a red bill. Males have a black chin and a black-bordered, rose-coloured neck-ring.

Vocalisations: a high and harsh *kyee-kyee-kyee-kyee*.

Where to see: this introduced species breeds on Barbados and Martinique.

Yellow-bellied Elaenia

Elaenia flavogaster
L 17cm WS 20–22cm

Very similar in appearance to the regionally more widespread Caribbean Elaenia (next page), but with a thicker, more obvious olive-brown crest often held fully erect. The vent and belly are bright yellow tinging the grey breast. The throat is grey. The upperparts and wings are dull olive-brown and there is a white double wing-bar. The long, slender brown tail is notched.

Vocalisations: like Caribbean but more explosive.

Where to see: dry scrub and secondary forests and gardens on St Vincent and Grenada.

Caribbean Elaenia *Elaenia martinica* L 17cm WS 20–22cm

A typical tyrant flycatcher, with a long, slender body and notched tail, olive-brown upperparts and pale grey underparts (with a faint yellow to the belly and vent). Two pale wing-bars and tawny streaking break up the olive-brown wing. An olive-brown crest is often held semi-erect. The bill is brown above and pale horn below.

Vocalisations: a single high-pitched, descending whistle – *pshREEee*.

Where to see: throughout the region in mixed woodland, dry scrub and forest edges.

Tropical Kingbird *Tyrannus melancholicus* L 20cm WS 38–41cm

Similar to Grey Kingbird (overleaf), but slightly smaller with bright yellow underparts and olive-grey upperparts. Both have a grey crown and face, but Tropical has a smaller bill and less obvious mask. Note the white throat.

Vocalisations: a series of twittering peeps and trills.

Where to see: has bred on Grenada in open grassland and dry scrub.

Grey Kingbird *Tyrannus dominicensis* L 23cm WS 37–41cm

A starling-sized flycatcher with medium grey upperparts, whitish underparts and grey shading on the breast. The dark greyish-brown wings show two faded wing-bars and prominent pale edges to the feathers. The large head shows a dull grey crown and dark eye-stripe, creating a heavily masked appearance. The thick black bill is slightly hooked at the tip, and the dark tail is notched. Often perches in plain view on telephone wires or bare branches.

Vocalisations: a high-pitched and rolling whistle – *ptrEEE-ptrEEE*.

Where to see: common in all habitats.

juv.

Fork-tailed Flycatcher *Tyrannus savana*

L 35cm including tail WS 38–39cm

Similar plumage to Grey Kingbird but has spectacularly long black tail-streamers more than twice the length of its body. There is a glossy, black cap covering the eye. Females and juveniles are duller with shorter tail-streamers.

Vocalisations: twittering peeps and a buzzy flat *tk-tk-tk-tk*.

Where to see: summer on Grenada, where small flocks can occasionally be seen in open grassland.

Lesser Antillean Flycatcher *Myiarchus oberi* L 21cm WS 40–42

Similar size and shape to the more regionally widespread Grey Kingbird, but with brown upperparts, a greyish-brown head and an often fluffed-out brown crown. The throat and breast are grey, the belly and vent a soft, lemon-yellow. Note the two pale wingbars, along with an obvious and heavy presence of rufous-chestnut in the wing and in the lengthy notched tail. The heavy bill is black and slightly hooked. The size and colour of the endemic and near-endemic subspecies found on various islands can vary substantially. On St Lucia, for example, *sanctaeluciae* is large and has much rufous in the tail; the smaller *sclateri* on neighbouring Martinique has none.

Vocalisations: a high-pitched *schreeee-eee-eee-chreee.*

Where to see: endemic to the Lesser Antilles and possible across all habitats from St Lucia, north.

Grenada Flycatcher *Myiarchus nugator* L 20cm WS 38–41cm

Similar in size and shape to Grey Kingbird, but with a brown head and upperparts and an often fluffed-out brown crown. The throat and breast are a soft grey, the belly and vent lemon-yellow. Note the two pale wingbars and rufous-chestnut panelling in the wings and long notched tail. Typically shows a pink-tinged mandible. Often perches on exposed branches in dry scrub and at forest edges.

Vocalisations: a jerky, staggered *preep-preep-preep-preep-preep … preepprtruuuu.*

Where to see: a near-endemic found on Grenada and St Vincent (more reliable and plentiful on the latter).

Lesser Antillean Pewee *Contopus latirostris* L 15cm WS 22–25cm

Small flycatcher with a relatively large, olive-brown and slightly crested head, olive-brown upperparts, and subtly notched tail. Underparts are a warm beige though a distinctive subspecies on St Lucia shows rich peachy orange underparts. The throat is pale and dark wings show subtle cream streaking but, crucially, no wing-bars.

Vocalisations: flat first note followed by buzzy second.

Where to see: St Lucia – where there is a distinctive subspecies – Guadeloupe, Dominica, and Martinique. Primarily seen in mature forests and along forest edge.

Red-eyed Vireo *Vireo olivaceus* L 15cm WS 25–26cm

Similar to the regionally more common Black-whiskered Vireo (below), but with white underparts, a shorter bill, deeper red eye and no malar stripe. The paler eyebrow of the Red-eyed is bordered above and below with a thin black crown line and a black eye-stripe.

Vocalisations: a series of short undulating whistles.

Where to see: an uncommon autumn migrant seen in canopy of mixed deciduous woodland on northern islands.

Black-whiskered Vireo *Vireo altiloquus* L 16cm WS 25cm

A plain species with a dull, greyish-brown crown and olive-brown upperparts. The underparts are cream, the vent and flanks yellow-tinged. The olive-grey face is highlighted by a cream-coloured eyebrow and dark stripe running through a brownish-red eye. A thin black malar stripe descends diagonally from the base of the dark bill.

Vocalisations: a monotonous, frequently repeated double-noted *chuhp-chuhp*.

Where to see: wooded areas throughout the region, especially on Barbados, St Vincent and St Lucia.

Yellow-throated Vireo *Vireo flavifrons* L 14cm WS 23cm

A slightly smaller but large-headed and heavy-billed vireo, with an olive-green crown and face highlighted by a pair of yellow 'spectacles' extending from the base of the bill to encircle the eye. The upperparts are olive-green, becoming progressively greyer towards the rump. The throat and breast are brilliant yellow, and two white wing-bars show on dark grey wings. The belly and vent are greyish-white.

Vocalisations: a *whit* contact call. Usually forages silently on its own.

Where to see: rare, but recorded with some regularity in woodlands on Guadeloupe.

Cliff Swallow *Petrochelidon pyrrhonota* L 14cm WS 25–27cm

A squat, chunky swallow with a steel-blue crown and upperparts that are punctuated by a cream collar, vertical white streaking on the back and a chestnut rump. The underparts are creamy-white, the wings and square tail a faded bluish-brown. Distinguished from other swallows in the region by its pale cream forehead and the black patch on an otherwise rich rufous throat.

Vocalisations: a grating mix of high-pitched squeaks and twitters interspersed with a harsh, skittering *chrrr*.

Where to see: seen primarily over wetlands and open grassland on southern islands from September–December.

Cave Swallow *Petrochelidon fulva* L 13cm WS 23–25cm

Almost identical to Cliff Swallow, but slightly smaller, with a rich chestnut-orange forehead and no black on the throat.

Vocalisations: a confused series of staccato *cheep*, *kek* and peeping whistles.

Where to see: an uncommon migrant most often seen over open grassland close to water and sea cliffs.

Barn Swallow *Hirundo rustica* L 18cm WS 34cm

The 'classic' swallow, with a beautiful steel-blue crown and upperparts contrasting with black wings, pale to buff-orange underparts and a brilliant rufous-orange chin, throat and forehead. The tail has long, elegant, dark tail-streamers (absent from all other swallows in the region), and a spotted white terminal band.

Vocalisations: a muddled series of squawky undulating chirps, tweets and whistles, punctuated by rattling clicks.

Where to see: the most frequent migrant swallow in the region, seen regularly over grasslands from August to October.

Bank Swallow *Riparia riparia* L 14cm WS 25–33cm

A plain swallow with a pale brown crown and upperparts, white underparts and a single, pale brown breast-band. The tail is subtly forked.

Vocalisations: a rapid, irregular and coarse-sounding chatter of *chik*, *cherr* and *chee* calls.

Where to see: a migrant mostly to the southern islands, usually seen near mangroves, lagoons, sand dunes and other coastal sites.

Caribbean Martin *Progne dominicensis* L 19 cm WS 38cm

The largest swallow in the region. The male's head and upperparts, throat, breast and flanks glisten with a dark, metallic, purple-blue. The female has less sheen to her dark head and upperparts, and sports a greyish-brown throat, breast and flanks. The belly and vent are white in both sexes and the tail forked.

Vocalisations: an explosive mix of short, sharp, chirps and *preep* notes.

Where to see: common between March and October, when it breeds throughout. Frequently seen near sea cliffs or over grasslands and mangroves.

♂

House Wren *Troglodytes aedon* L 12cm WS 15cm

A tiny bird with a lengthy, slightly decurved bill (darker above and paler below), a faint, pale eyebrow, variable dark, thin barring on the lower back, wings and tail, and pale brown legs. Depending on subspecies, the upperparts (always darker than the underparts) range from a rich rufous-brown to beige, while the underparts can be dull chestnut-brown to white.

Vocalisations: typically very musical, flittering and scratchy, but varying between subspecies.

Where to see: endemic subspecies are known in forests on Dominica, St Lucia, St Vincent and Grenada.

White-breasted Thrasher *Ramphocinclus brachyurus* L 22cm WS 26–29cm

A grackle-sized species with plumage almost perfectly divided. The face, crown, upperparts and wings are black (or dark brown on Martinique), while the chin, throat, breast and belly are white. The flanks show dark shading and the tail and vent are dark. Juveniles have completely brown underparts. Often forages among vegetation near the ground in small family groups.

Vocalisations: a raspy *crraaaghhhh-eeeyaghhh-aghhhh.*

Where to see: a near-endemic frequenting dry scrubland and dry broadleaved forests on the south-eastern coast of St Lucia and eastern Martinique (especially Presqui'ile de la Caravelle).

Scaly-breasted Thrasher *Allenia fusca* L 23cm WS 27–30cm

Slightly smaller and more slender-bodied than Pearly-eyed (overleaf), with darker brown upperparts, a pale lemon-yellow eye and wings that show an intermittent, single, thin, white, wing-bar. The pale cream underparts are heavily dappled with brown, giving the throat, breast and flanks a densely scaled appearance. The central belly and vent are largely unmarked, and the long brown tail is tipped with white.

Vocalisations: shrill whistles and lower-pitched chirps.

Where to see: a Lesser Antillean endemic prevalent in forest edges and in woods on most islands.

juv.

Pearly-eyed Thrasher *Margarops fuscatus* L 26cm WS 30–32cm

Slightly larger than a grackle, with a distinctive, pale, pearl-coloured eye and a large, creamy-yellow bill. The head and upperparts are medium brown, while the underparts are cream, punctuated with intricate brown V-shaped chevrons, becoming sparser and hence more distinguishable on the flanks and the vent. The central belly is largely unmarked. The long, brown tail is tipped with white.

Vocalisations: mixed chirps, tweets and whistles.

Where to see: woodland and forested habitat from St Lucia to Anguilla.

Brown Trembler *Cinclocerthia ruficauda* L 24–26cm WS 29–32cm

Rufous with a grey-tinged head, glaring yellow-orange eye and a long, black, slightly decurved bill. The breast can show grey shading, and wings and tail carry a chestnut hue. The tail is often cocked. It regularly performs a remarkable, whole-body, quivering display. Arboreal, but occasionally forages on the ground.

Vocalisations: rolling and undulating warbled whistles.

Where to see: a Lesser Antillean endemic found on St Vincent, Grenada, Guadeloupe, Dominica, Montserrat, Barbuda, St Kitts and Nevis, and Martinique, in dense tropical forest, secondary growth and along forest edge.

Grey Trembler *Cinclocerthia gutturalis* L 26cm WS 30–32cm

This lengthy mimid has a dark brown-grey crown, upperparts and wings, along with paler ashy-beige underparts (heavily shaded on the breast and flanks). The face has a dark grey mask – perfectly encapsulating a bold, amber-coloured eye – and an extremely long, black, decurved bill. It often performs an identical quivering display to Brown Trembler (previous page).

Vocalisations: alternating high- and low-pitched whistles.

Where to see: a near-endemic to St Lucia and Martinique, where it inhabits dense primary forests, dry scrubland and orchards.

Tropical Mockingbird *Mimus gilvus* L 24 cm WS 35cm

A long, slender species with a grey head and upperparts, a black eye-stripe and broad white eyebrow, whitish underparts and a long, slender, dark grey tail tipped with white. The shortish, black bill is slightly decurved and the black legs are long. There is a significant amount of black in the folded wing. Juveniles have brownish-grey upperparts and streaked breasts.

Vocalisations: a varied series of sharp, high-pitched whistles, peeps and trills.

Where to see: absent from Barbados and islands north of Antigua and Barbuda, but common elsewhere in most habitats, including residential areas.

juv.

Rufous-throated Solitaire *Myadestes genibarbis* L 20cm WS 38–40cm

A beautiful, 'fairy-like', forest-dwelling thrush. The head and upperparts are bluish-grey (black on St Vincent), while the greyish underparts are highlighted by an orange throat patch and vent. There is a cream-coloured crescent under the eye, cream gape and chin, and a black malar stripe. Juveniles show orange-flecked upperparts and mottled tawny-orange below.

Vocalisations: the haunting and ethereal whistling song consists of two identical notes followed by a higher and a lower one.

Where to see: a breeding resident on St Lucia, St Vincent, Dominica and Martinique in dense forest.

Cocoa Thrush *Turdus fumigatus* L 21.5–24cm WS 36–37cm

A medium-sized, all-brown thrush. The upperparts are a slightly darker rufous-brown, while the underparts show a faint chestnut hue. The throat is paler and streaked, the vent cream and the bill and legs brown.

Vocalisations: a regular, prolonged, whistled, two-note chirp, often part of the dawn chorus.

Where to see: St Vincent and Grenada in forests and residential gardens.

Spectacled Thrush *Turdus nudigenis* L 24cm WS 34–37cm

A heavy-set thrush with a brown head and upperparts, an obvious, bright yellow eye-ring and a yellow bill. The underparts are dusky-beige with a paler belly and vent. The pale throat is streaked brown, and the long dark tail is notched. Juveniles show faintly streaked orange upperparts and a dappled breast.

Vocalisations: melodic trilled whistles often culminating in two distinct tweets.

Where to see: absent from Barbados but otherwise found from Guadeloupe to Grenada in woodland and residential areas.

Forest Thrush *Turdus lherminieri* L 28–30cm WS 42cm

A large thrush with a warm chocolate-coloured head and upperparts, a thick, yellow eye-ring, and a yellow bill and legs. The throat, breast, belly and vent are white, and heavily chevroned with brown, giving the appearance of underparts festooned with brown-trimmed white arrowheads. The extent of the chevrons varies with each island subspecies.

Vocalisations: a melodious, undulating *truup-uh-truup-tchreeee*.

Where to see: a secretive forest-dweller and Lesser Antillean endemic, known only on Guadeloupe, Montserrat, St Lucia and Dominica. Best seen at dawn in the Basse-Terre Mountain Range, Guadeloupe.

Red-legged Thrush *Turdus plumbeus* L 27cm WS 40–41cm

Strikingly patterned, large, long-tailed thrush, with ash-grey plumage highlighted by a black mask and a reddish-orange eye-ring, a bright orange bill, white throat streaked vertically with black lines and dazzling reddish-orange legs. The upperwing is streaked black, while the belly, vent and undertail are white. No throat streaks in juveniles.

Vocalisations: a mix of short, sharp cheeps and *prrrp* calls.

Where to see: found on Dominica in residential gardens and along forested roadside verges, especially at dawn.

Northern Red Bishop *Euplectes franciscanus* L 11cm WS 19–22cm

A brightly coloured, tiny African species. The male is unmistakable with brilliant reddish-orange plumage, a black face, crown and belly, and brown wings. The female has a brownish-black mottled crown and upperparts, pale beige underparts (with fine streaking on breast), a cream-coloured eyebrow, brown eye-stripe and a pale bill.

Vocalisations: a high-pitched tweeted *teet-teet-teet-teet*.

Where to see: introduced to Martinique and Guadeloupe and often seen in flocks along the edges of fields, grass verges and parks.

♀

♂

Village Weaver *Ploceus cucullatus* L 17cm WS 24–25cm

Males are bright yellow with a jet-black head and throat, rufous-maroon nape and a gleaming red eye. The wings and back show significant black, and the large black bill is sharply pointed. Females are dull olive-yellow and show brown on the back and wings. The bill and legs are pale. Nests in noisy colonies in unique hanging nests.

Vocalisations: a confused chatter of whistles and *screek* notes.

Where to see: introduced to Martinique (and known on Guadeloupe). Found in dry scrubby vegetation, mixed woodland and grassland.

♂

♂ non-br.

Orange-cheeked Waxbill *Estrilda melpoda* L 10cm WS 17–19cm

A tiny bird with a silver-grey crown and underparts and orange belly patch, bright orange cheeks, a heavy red bill (black in juveniles) and a long, dark brown tail, the base of which is red. The back and wings are brown.

Vocalisations: a subtle, high-pitched *tsiiii*.

Where to see: introduced to Guadeloupe and Martinique, where it favours tall, mature grasslands and scrub.

Black-rumped Waxbill *Estrilda troglodytes* L 10cm WS 17–19cm

A tiny species, with soft, greyish-beige plumage, a brilliant bright red bill and a broad red eye-stripe extending to the nape. Note the black rump, long black tail and white vent in both sexes. Males show a faintly barred, rose-tinted breast, belly and flanks. Juveniles show a black eye-stripe and black bill.

Vocalisations: alternating *chuup* and *thhwee* notes.

Where to see: introduced to Martinique and Guadeloupe, where it favours large grasslands and swamp edges.

Common Waxbill *Estrilda astrild* L 11cm WS 12–14cm

Similar to Black-rumped Waxbill but has darker grey, finely barred plumage, shorter wings, a greyish-brown rump and tail, black vent, and a rose-coloured patch on the paler breast and belly. The bill and broad eye-stripe are red in adults, while juveniles show a narrow eye-stripe and black bill.

Vocalisations: mixed *chikituh tseeees chikituh tseeeees.*

Where to see: introduced to Martinique, where it favours grassy areas, often close to residential neighbourhoods.

Red Avadavat *Amandava amandava* L 10cm WS 11–14cm

A very tiny bird. Breeding males are unmistakably vibrant red with matching red bills. A smattering of tiny white dots covers the neck, flanks and the otherwise dark brown wings. Females and non-breeding males are grey with red bills and rumps, and have dark, white-dotted wings. Both sexes show a dark eye-stripe and tail and pinkish-cream legs.

Vocalisations: high-pitched trills, cheeps and descending *tsiiiuu* notes.

Where to see: introduced to Guadeloupe and Martinique, where it can be found in grassland.

Scaly-breasted Munia *Lonchura punctulata* L 11cm WS 20–25cm

A small, spectacularly patterned species. The breast and flanks show dense brownish-black scales, while the remaining underparts are an unscaled cream. Note the deep, rich chestnut hood, dark throat and rufous-brown upperparts. The bill is black and legs dark grey. Juveniles have dark tawny upperparts and paler unmarked underparts.

Vocalisations: a short sharp *preet.*

Where to see: introduced on Martinique, St Kitts, Antigua, Guadeloupe and Dominica, where it can be seen in parks, grassland areas and along roadsides.

Chestnut Munia *Lonchura atricapilla* L 11cm WS 10–15cm

This tiny species has a rich chestnut plumage and a full black 'executioner's' hood descending to the upper breast. Note also the thick black patch running from the vent to the belly. It has a disproportionately large, blue-grey bill and legs. Juveniles are dark chestnut above and paler below.

Vocalisations: a short, two-note whistle.

Where to see: introduced to Martinique, where it favours marshes and grassland (especially near the coast).

White-headed Munia *Lonchura maja* L 11cm WS 12–14cm

Tiny species with an obvious creamy-white 'executioner's' hood, warm chestnut body and a very pale bluish-grey, oversized bill. Juveniles have a brown head, pale chin and throat, and a greyish-black bill.

Vocalisations: a lengthy *wheeeee … phee … heee.*

Where to see: introduced to Martinique, where it favours low-lying grasslands and marshes.

House Sparrow *Passer domesticus* L 16cm WS 24cm

Males are patterned cinnamon-brown and black on the back and wings, have a grey crown and breast, and a black eye-mask, bib and chin. Females have a beige-and-brown mottled back and wings, beige breast and a pale eyebrow. Both sexes show a single pale wing-bar.

Vocalisations: a repeated two-noted *chyuhp-choop.*

Where to see: a long-established introduced species associated with human habitation on Antigua, Guadeloupe and St Vincent.

♂

♀

Lesser Antillean Euphonia *Euphonia flavifrons* L 10cm WS 12cm

Tiny bird with a striking turquoise-blue crown and nape, dark cheeks and a yellowish-orange forehead and throat. The upperparts are olive-green (darker on the wings and tail) and the rump yellow. The male's underparts vary from olive-yellow to yellowish-orange, while the female's are typically a dull yellowish-green.

Vocalisations: a very high-pitched *tseeeeee*.

Where to see: a Lesser Antillean endemic seen on most islands but notably absent from Barbados. Inhabits moist forest canopies at high elevations, where it is associated with Mistletoe Vine (*Psittacanthus martinicensis*) berries.

Bobolink *Dolichonyx oryzivorus* L 17cm WS 27cm

A finch-sized species with a flatter head. Breeding males are jet-black with a cream nape and silver-grey rump. The back and wings show heavy greyish-white and buff streaking, and the long tail is sharply serrated at the tip. The female is tawny-buff with thick brown streaking on the back and wings, a dark crown with a pale central stripe, pale cream eyebrow, partial black eye-stripe and a pale pink bill and legs.

Vocalisations: a wide repertoire of high-pitched squeaks and chips.

Where to see: a scarce migrant to open grasslands on Barbados and Guadeloupe.

♀

♂

Martinique Oriole *Icterus bonana* L 19cm WS 30cm

A striking oriole with a deep auburn head, neck and breast, and rich orange belly, vent and rump. A brilliant orange patch shows on the shoulders of the folded wings, which are otherwise jet-black, as is the back and long tail (serrated at tip). The sharply pointed bill is silver and black, and the legs silver-grey.

Vocalisations: a whistled *'hey you there'*.

Where to see: endemic to Martinique and best seen along the montane forest trail to Piton Boucher. Fond of *Heliconia* stands.

Montserrat Oriole *Icterus oberi* L 21cm WS 28–30cm

A slim, starling-sized species. Males are jet-black with a rich yellowish-orange belly, vent and rump. Females and juveniles have mustard-olive plumage, typically darker on back and wings. The pointed bill is black above and silver below, and the long tail is slightly serrated at the tip.

Vocalisations: a series of high-low 'swinging' whistles.

Where to see: endemic to Montserrat, where regularly seen on the Oriole Trail. Prefers moist tropical forest, where it is often seen in *Heliconia* stands.

♀

♂

St Lucia Oriole *Icterus laudabilis* L 21cm WS 28–30cm

Stunning starling-sized bird with shiny black plumage punctuated with dazzling orange-yellow patches on the shoulders of the folded wings, and an orange-yellow lower back, rump, belly, belly flanks, vent and thighs. The bill is silver-grey at the base with a black tip. Juveniles are brownish-cinnamon, with a black face, throat and tail, and two pale wing-bars.

Vocalisations: a harsh, single-noted *ehrrgh*.

Where to see: endemic to St Lucia in wooded areas.

Carib Grackle *Quiscalus lugubris* L 23cm 32–35cm

The piercing yellow eye of adults is diagnostic. Males are glossy black, while females vary from pale grey to black depending on the island and subspecies. Juveniles are brown with a black eye on most islands. Often seen in mixed flocks with Shiny Cowbird.

Vocalisations: has a squeaky, 'high-low-high' whistled song and single *chuhp* call.

Where to see: found throughout the region in all habitats.

♂

♀ St Lucia

♀ Barbados

Shiny Cowbird *Molothrus bonariensis* L 19cm WS 29–30cm

A slender, 'streamlined' species with glossy black adult males that show purple iridescence in favourable light. Females are brown (darkest on the wing). The eyes and stubby, sharp bill are black. Has a diagnostic undulating flight style. Juveniles have a yellowish eyebrow, brownish upperparts and tawny-yellow, streaked underparts.

Vocalisations: fast high-pitched whistles, tweets and rolling trills. Males produce a soft repeated grunt during courtship.

Where to see: possible throughout the region in a variety of habitats.

♂

♀

Ovenbird *Seiurus aurocapilla* L 15cm WS 24cm

Like the more frequent Northern Waterthrush, with darkish upperparts and pale, heavily streaked underparts. The upperparts are olive-green, the underparts streaked with black. There is a bright orange, black-bordered central crown-stripe and a prominent white eye-ring on the olive head. Forages among forest leaf litter.

Vocalisations: a hard *chip* and a *sweeet* call in flight.

Where to see: an uncommon migrant in dense forest, woodland and mangroves on all islands.

Louisiana Waterthrush *Parkesia motacilla* L 15cm WS 25.5 cm

Almost identical to the more frequently sighted Northern Waterthrush, but larger, with whiter underparts showing less streaking (and none on the throat). The long eyebrow is whiter and broadens behind the eye. Constantly bobs tail.

Vocalisations: like Northern (below) but quieter and less metallic.

Where to see: scarce but possible close to water on any island with well-forested habitat.

Northern Waterthrush *Parkesia noveboracensis* L 13cm WS 24cm

Identical male and female show warm chocolate-brown upperparts and creamy underparts that are heavily streaked brown. The largely brown face and head show a long cream-coloured eyebrow that tapers towards the nape, and a pinkish-brown bill. Legs are a soft pink. Constantly bobs rump and tail up and down.

Vocalisations: a unique, metallic *chenk* … *chenk* and a *zzipp* flight call.

Where to see: a common migrant to the lower storeys of mangrove swamps.

Black-and-White Warbler *Mniotilta varia* L 13cm WS 21cm

Unmistakable with its spectacular black-and-white streaked plumage. Skulks horizontally and vertically along branches and trunks like a treecreeper.

Vocalisations: a sharp rattle and a hissing *fssss* flight call.

Where to see: regular migrant throughout to forested areas (especially on Guadeloupe, Martinique and Dominica).

♂

♀

Prothonotary Warbler *Protonotaria citrea* L 14cm WS 22–23cm

Breeding male has a stunning mango-yellow head, throat, breast and belly, with soft, bluish-grey wings and tail. The back is yellowish-green and the vent white. Females are similar but duller. The species' eyes are glossy black and the legs dark grey.

Vocalisations: simple *cheep-cheep-chEEP-CHEEP.*

Where to see: regular migrant in woodland throughout.

Kentucky Warbler *Geothlypis formosa* L 13cm WS 21cm

Olive-green above and bright yellow below, this species' head is particularly striking, with its thick golden-yellow eyebrow and partial eye-ring contrasting with a prominent black mask. Females are similar but duller. The bill is dark and the legs pale pink.

Vocalisations: a low *chock* call, and a buzzing *drrrtt* in flight.

Where to see: a rare but regular migrant to the mangrove forests of Guadeloupe and islands further north.

♀

♂

Common Yellowthroat *Geothlypis trichas* L 12cm WS 17cm

A small warbler with primarily olive-green plumage, highlighted in breeding males by a dazzling yellow throat, breast and vent, and a broad black mask (with a thick grey border above). Females are similar but duller, and have a greyish-brown face and crown, fainter eyebrow and pale eye-ring.

Vocalisations: a dry *jek* and a buzzing *dzik* in flight.

Where to see: uncommon migrant throughout to moist forested habitat, foraging close to the ground.

♂

♀

Whistling Warbler *Catharopeza bishopi* L 14 cm WS 21–22cm

A sooty-black warbler with a prominent white eye-ring and greyish-white underparts, punctuated by a black breast-band and dark flanks. Often holds its head and long tail cocked in a pronounced V-shape – a distinguishing feature of this species. The legs are pale and the bill black. Juveniles are brown and also cock their tails. Shy and retiring.

Vocalisations: a continuous, rapid-fire, punctuated whistle increasing in intensity and length.

Where to see: endemic to St Vincent, where it favours moist montane forest. Look for this bird in tangled vegetation along the La Soufrière Trail

Plumbeous Warbler

Setophaga plumbea L 13cm WS 21cm

A mostly grey warbler with a long, thick, white eyebrow, white under eye-crescent and two thin white wing-bars. The upperparts are darker, while the underparts are greyish-white with dark grey flanks. A long, dark grey tail is tipped white and often flicked. Juveniles are greyish-olive above, yellowish-grey below and show a yellow eyebrow.

Vocalisations: a series of sharp whistled undulating notes – *phee-twii-tiuh … tee-tee-tiuh.*

Where to see: endemic to Dominica and Guadeloupe where it favours dense forest undergrowth and the lower storeys.

Hooded Warbler *Setophaga citrina* L 13cm WS 17–18cm

Another olive above, yellow below, warbler species in which the adult male has a black hood enclosing a broad, bright yellow mask. It shows white on the underside of a tail that is flicked constantly. Females and immature males have a faint black outline framing the yellow face and an olive-green crown. Both sexes show dark lores.

Vocalisations: a flat squeaky *tiiip*, more ascending in flight.

Where to see: a rare migrant seen primarily in the mangroves and tropical forests of the northern islands of Anguilla, St Kitts and Sint Maarten.

♀

♂

American Redstart *Setophaga ruticilla* L 12cm WS 19–20cm

The breeding male is a striking black with bright orange patches on the upper flanks, wings and tail, and an off-white belly and vent. The female has a grey head, split white eye-ring, olive-green upperparts and grey-white underparts, with pale yellow patches as in the male.

Vocalisations: a high, squeaky *chip*, plus a squeaky *tsweeet* in flight.

Where to see: a common migrant throughout to all wooded habitats.

♀

♂

Cape May Warbler *Setophaga tigrina* L 13cm WS 21cm

The breeding male's olive-green upperparts and yellow underparts are streaked with black, contrasting with a thick yellow unmarked collar. The crown is dark and the face yellow, with a large orange patch surrounding the eye (grey in females) and a white patch on the upperwing (two pale wing-bars in females). Both sexes show a yellow rump.

Vocalisations: a high, hard *tsii*, plus a buzzing *tzeww* in flight.

Where to see: an uncommon migrant to wooded areas and gardens on northern islands, including Saint Martin and Anguilla.

Northern Parula *Setophaga americana* L 11cm WS 17–18cm

A very small, multi-coloured warbler with two white wing-bars. Breeding males show a bright grey-blue crown and upperparts with a large olive-yellow patch on the upper back, along with a bright yellow, black-bordered throat and a breast topped with an untidy rufous breast-band. The belly and vent are white. Females are similar but duller and typically lack a breast-band.

♀

Vocalisations: a very loud *chip*, along with a high repeated *tsiipp* flight call.

Where to see: a common autumn migrant to forests on Guadeloupe and islands further north.

♂

Magnolia Warbler *Setophaga magnolia* L 12cm WS 19–20cm

Breeding males have a grey crown and black upperparts, contrasting with bright yellow underparts (the flanks and breast streaked black) and a white vent. A large white patch is visible in the dark wing. A white eyebrow and crescent below the eye contrast with a black mask. The breeding female is similar but paler and shows two pale wing-bars and a lighter, olive-green back. All plumages show a yellow rump, a broad black terminal band and a white undertail.

♂ non-br.

Vocalisations: a distinctive *chwiff* plus a soft, trilling *zip* flight call.

Where to see: rare migrant to woods on northern islands.

♂

Yellow Warbler *Setophaga petechia – aestiva* group L 12cm WS 20.5cm

Breeding males have a bright yellow head and underparts with rufous streaking on the breast and flanks. Some orange-rufous shading is present on the crown. Breeding females show duller yellow underparts with very faint rufous streaking. The upperparts of both sexes are olive-green with black streaks on the darker olive wings. Non-breeders are duller overall.

Vocalisations: a loud *chip* call plus a trilling *zipp* in flight.

Where to see: an occasional migrant to the region and possible in any habitat.

Yellow Warbler *Setophaga petechia – petechia* group L 13cm WS 21cm

Breeding males show a bright yellow body, rufous streaking on the breast and flanks, and have rufous on the head (the extent of which varies depending on the subspecies). The upperparts are yellowish-olive with black in the wing. Females resemble Yellow Warbler but with no rufous streaking on the underparts.

Vocalisations: a sharp musical song – *shree-shree-shree-shree-see-see-see-see.*

Where to see: the breeding resident Yellow Warbler subspecies found throughout the region in a wide range of wooded and scrub habitats.

♀

♂

Chestnut-sided Warbler *Setophaga pensylvanica* L 12cm WS 19–20cm

Smartly patterned, with breeding males sporting a bright yellow crown, white face with black mask and 'moustache', and white underparts with rich chestnut flanks. The black back and wings are tinged with olive, and there are two cream wing-bars. Breeding females are slightly duller with a faded black mask. Non-breeders lack mask, show a pale eye-ring and have yellow-green upperparts.

Vocalisations: *chip* calls with a buzzing *jrrt* in flight.

Where to see: regular migrant to woodlands on Guadeloupe; otherwise uncommon.

♂ non-br.

♂

Blackpoll Warbler *Setophaga striata* L 14cm WS 22–23cm

Non-breeders have dark-streaked, dusky-olive upperparts and cream underparts with faint grey flank streaking. Note two white wing-bars on the olive wings and the pale eyebrow. Breeding males have black caps, large white cheeks, a thick black malar stripe and white underparts, with bold black flank streaks. A white vent, pale-brown bill and orange-yellow legs are consistent in all plumages.

♀

Vocalisations: a neat *chip* and a buzzing flight call.

Where to see: a common migrant to sparse woodland near fresh water.

♂

Black-throated Blue Warbler *Setophaga caerulescens*

L 13cm WS 19–20cm

Breeding males are deep blue with a black face, throat and flanks, and bright white underparts. Females are greyish-olive on the crown and upperparts, with greyish-beige underparts and a long cream-coloured eyebrow. A small white patch shows on the wing in all plumages. The bill is dark and the legs pale.

♀

Vocalisations: a very high-pitched *stipp*, plus a sharp *twik* flight call.

Where to see: a rare migrant throughout to dense moist forests.

♂

Palm Warbler *Setophaga palmarum* L 13cm WS 20–21cm

Breeding birds show greyish-brown upperparts and yellow underparts, rufous streaking on the breast-sides and flanks, and a rufous crown above a well-defined, yellow eyebrow. Non-breeders are much duller with a greyish-brown crown, pale yellow eyebrow and much-reduced streaking to dull yellow underparts. Bobs its tail incessantly and generally stays near ground level.

Vocalisations: a sharp, husky *tchip* and a *tsink* flight call. Keeps near ground level.

Where to see: scarce migrant to scrub and dry woodland.

br.

non-br.

Prairie Warbler *Setophaga discolor* L 12cm WS 17–18cm

Adult males have an olive-green crown and upperparts with thick maroon streaks on the back and two yellow wing-bars on greyish-olive wings. Underparts are bright yellow and the flanks are heavily streaked with black. Note the thick yellow eyebrow, thin black eye-stripe and yellow crescent under the eye (bordered with black below). Adult females are paler and duller.

Vocalisations: similar calls to Palm but less husky.

Where to see: an uncommon migrant most often sighted on northern islands such as Anguilla, Guadeloupe and Montserrat. Favours dry mixed woodland, scrub and mangroves.

Barbuda Warbler *Setophaga subita* L 13cm WS 19–20cm

Grey upperparts contrast with bright yellow underparts, with a white vent and partially white undertail. There are two white wing-bars on the greyish-brown wings. The crown and face are grey, with a bright yellow eyebrow, a black eye-stripe and a thick yellow crescent beneath the eye (bordered with black below). Juveniles are brownish-grey above and duller below.

Vocalisations: a long, high-pitched trill.

Where to see: endemic to Barbuda, where it favours dry scrubland.

Canada Warbler *Cardellina canadensis* L 13cm WS 20–21cm

Smooth slate-blue upperparts contrast with stunning yellow underparts and thick, glossy black necklace streaking in the male. The female is similar but duller. Both sexes show yellow 'spectacles'.

Vocalisations: a dry, sharp *tiup*, plus a slurred *chwit* in flight.

Where to see: rare throughout in moist forests.

St Lucia Warbler *Setophaga delicata* L 14cm WS 20–21cm

The rich blue-grey crown and upperparts contrast with a brilliant sunshine-yellow throat, breast and belly, while the vent and undertail are white. The boldly marked face features a heavy yellow eyebrow bordered above by a thin black line, black eye-stripe and a thick yellow crescent under the eye, bordered below by a thin black line. Two white wing-bars complete the ensemble. Juveniles are brownish-grey above and duller below.

Vocalisations: an undulating warbled trill.

Where to see: a common St Lucia endemic in all vegetated habitats.

Black-throated Green Warbler *Setophaga virens* L 13cm WS 19–20cm

Males show an olive-green crown and upperparts, dark grey wings with two white wing-bars, a bright yellow face with olive-green ear-coverts, a black throat, flank streaks and upper breast, and white underparts. Outertail feathers white and vent yellow. Females are similar but paler, with a creamy yellow throat and grey streaking on underparts.

Vocalisations: a high *swit* in flight; otherwise, a sharp *teck*.

Where to see: a rare but regular migrant to broadleaved forests on Guadeloupe.

Rose-breasted Grosbeak *Pheucticus ludovicianus* L 19cm WS 32cm

A heavy-set, large-beaked finch, in which breeding males show a glossy black hood and upperparts and a white rump and lower back, black wings with large white patches and a bright red, Y-shaped patch on the white breast and belly. Females show dark-brown heads with a prominent off-white eyebrow, dark brown streaked upperparts, cream-coloured streaked underparts and brown wings with smaller white patches.

♀

Vocalisations: a soft, wheezing *wheeek* in flight; otherwise, a sound like a squeaky rubber sole.

Where to see: an uncommon migrant throughout. Favours broadleaved forests.

♂

Indigo Bunting *Passerina cyanea* L 15cm WS 20–21cm

The spectacular deep blue plumage of the breeding male is unlikely to be confused with any other species in the region. The blue on the wings and tail is interspersed with dark brownish-black feathering. The breeding female is tawny-brown above and pale beige below, with faint breast streaking and slight hints of blue in the wings and tail. The bill is horn-coloured and the legs black.

Vocalisations: a buzzing flight call and a dry, sharp *pwik*.

Where to see: uncommon solitary migrant throughout to fields or the woodland understorey.

♀

♂

Scarlet Tanager *Piranga olivacea* L 18cm WS 29–30cm

The breeding male is unmistakably bright red with black wings and tail. Females and non-breeding males are olive-green, differing in wing and tail colour (dull dark brown in females, black in males). Moulting males can show a mix of red, olive-green and brown. The pale bill and dark legs are consistent in all adult plumages.

Vocalisations: a two-note variation on *chik-brrr*, with a whistling *puwee* in flight.

Where to see: regular migrant to woodland, parks and gardens in the northern islands.

♂

♂ non-br.

Lesser Antillean Saltator

Saltator albicollis L 22cm WS 30–31cm

A slender grackle-sized bird with a 'tubular' appearance. The crown and upperparts are dull olive green, while the underparts are beige with faint olive-green breast and flank streaking. The head is olive-green with charcoal-grey cheeks and a long cream eyebrow. A black malar stripe contrasts against a white throat, and the disproportionately large grey beak shows a yellow tip and gape. The long tail is charcoal-grey.

Vocalisations: a piercing multi-noted descending and ascending *shrEE-she-shruh-seee-SheeEE*.

Where to see: a Lesser Antillean endemic found on Dominica, Guadeloupe, St Lucia and Martinique. Often sighted along forest edge and in pockets of dry sparse woodland.

Bananaquit *Coereba flaveola* L 10cm WS 19cm

Small, common black-and-yellow bird. Typically has a black head and upperparts with a yellow rump and bright yellow underparts with a grey throat and off-white vent, a prominent white eyebrow and a red gape at the base of the thin, black, decurved bill. Juveniles are duller with a yellow eyebrow. Plumage varies between island subspecies from subtle to significant (all-black plumage).

Vocalisations: a buzzing *tzuhheeeEE-EE-EEEE.*

Where to see: extremely common throughout and often frequents hotel grounds and restaurants, where incredibly 'acclimatised' to humans.

Barbados Bullfinch *Loxigilla barbadensis* L 15cm WS 20–21cm

Both sexes are almost identical in this unspectacular greyish-brown species. It is darker and browner on the crown, face and upperparts, paler and greyer on the underparts. There is a chestnut wing-panel, and the chin and vent are pale beige. The female is distinguished by a pale, horn-coloured mandible, while the male's entire bill is dark brown.

Vocalisations: a harsh, rapid *schipp-schipp-schipp*, followed by a slow *scherrrrrrp*.

Where to see: endemic to Barbados and known in every habitat. Very 'acclimatised' to hotel restaurants.

♀

♂

Lesser Antillean Bullfinch *Loxigilla noctis* L 15cm WS 20–21cm

The male is completely black with a rufous-orange throat and a small rufous-orange patch over the eye (subspecies on some islands also show a rufous-orange vent). The female is greyish-brown with a marginally darker and browner head and upperparts, and paler, greyer underparts, chestnut in the wing and a cream vent. The male's bill is black, the female's dark brown above and horn below.

Vocalisations: a harsh, repetitive *shcipp-schipp-schipp … tuck.*

Where to see: a common Lesser Antillean endemic known in most habitats throughout the region, but absent from Barbados and some of the northern islands.

♂

♂

♀

St Lucia Black Finch *Melanospiza richardsoni* L 13cm WS 18–20cm

Males of this small, warbler-sized species are jet-black, while females have grey hoods, rufous upperparts and wings, and lighter, beige-brown underparts. Both sexes have pink legs. The large conical beak is black in males and brownish grey in females. Often seen in pairs among leaf litter.

Vocalisations: a sharp *tsiii-tsiii-tsii … tsuhp*.

Where to see: endemic to St Lucia, where it forages in leaf litter in the dry forests of the east coast and dense low-lying foliage in mature forests such as Des Cartiers.

♂

♂

Black-faced Grassquit *Melanospiza bicolor* L 10–11cm WS 17cm

Adult males have a black head, the colour of which bleeds heavily downwards onto the throat and breast, while the remaining underparts show a soft greyish-olive. The upperparts and wings are dark olive-green in both sexes. Females show unmarked pale olive-grey underparts and heads. The dark brown, stubby, sharply pointed bill and pink legs are consistent to both sexes.

Vocalisations: a short, sharp, shrill *psziiii-tsup*.

Where to see: found throughout the region on roadside verges, fields and grassland.

Blue-black Grassquit *Volatinia jacarina* L 10cm WS 17cm

Breeding males are an iridescent bluish-black with white 'armpits' that flash rapidly during aerial displays. The females have brown upperparts and darkly streaked beige underparts. Juveniles recall females but show more heavily streaked underparts and darker wings and tails. The legs are a greyish-beige.

Vocalisations: a buzzing *tseeee.*

Where to see: during summer and autumn on Grenada's fields and grassland.

♀

♂

Yellow-bellied Seedeater *Sporophila nigricollis* L 9cm WS 15–16cm

Like the more regionally widespread Black-faced Grassquit, breeding males show black heads and throats and dark olive-green upperparts. Seedeater males, though, show a diagnostic lemon-yellow breast, belly and vent. Females of both species are similar, but leg colour is diagnostic: pink in the grassquit and black in the seedeater.

♂

Vocalisations: a rapid-fire, scribbly *whzee-whzee-whzee.*

Where to see: small flocks are common on Grenada's grasslands.

Chestnut-bellied Seed-finch *Sporophila angolensis* L 11cm WS 17–18cm

The male is black with a rich rufous-chestnut belly and vent, and an obvious small white wing-patch. The female is warm chestnut-brown, with darker wings and tail. Both sexes have disproportionately large, conical black bills.

♂

Vocalisations: a prolonged series of whistles and trills.

Where to see: introduced to Martinique, where it favours grasslands.

Grassland Yellow-finch *Sicalis luteola* L 10cm WS 17cm

A tiny, slender-bodied finch, with heavily streaked and checked brown-and-beige wings and upperparts, contrasting with unmarked bright yellow underparts (dull straw in females). The head is yellow but shaded greyish-brown, with the male showing more yellow in the face than the female and a prominent yellow eyebrow. The bill and legs are a pale greyish-horn, and the tail is notched. Males often sing from the top of bushes.

Vocalisations: an undulating series of ascending and descending whistles.

Where to see: savanna and open grassland throughout the region.

Lesser Antillean Tanager *Stilpnia cucullata* L 15cm WS 22–23cm

A slim House Sparrow-sized tanager. The male is buff with a chestnut crown (maroon on Grenada), prominent black mask and turquoise wings and tail. The underparts show a faint purplish tinge. The upper back of males is a blonde-yellow on Grenada. The female is similar but has greener wings and upperparts, and grey-tinged underparts. Juveniles show faded crowns.

Vocalisations: a series of high-pitched *tzsiiiips* followed by rapid, chaotic *tktktktktk*.

Where to see: a near-endemic known only to St Vincent and Grenada, each of which has its own distinct subspecies, occasionally accorded full-species status. Frequents woodland. Favours fruiting fig trees along the La Soufrière Trail, St Vincent.

GLOSSARY OF TERMS

Aigrettes Long, wispy, feathers growing from the nape or back of an egret.

Autumn migration The period spanning the months of July to October.

Breeding resident A species that breeds on a particular island and is present for all or a portion of the year.

Carpal bar A dark bar at the front edge of the wing.

Conspecific Of the same species

Crest Tousled or spiky crown feathering.

Crown The top of the head.

Decurved A bill that is significantly bent or angled towards its tip

Dry forest Mixed deciduous, usually sparse, wooded areas with low rainfall, shorter trees and a drought-tolerant flora.

Ear-coverts The feathers that cover the ears of a bird.

Endemic species A species that can only be found on a single island in the Lesser Antilles.

Eyebrow An area of feathering above the eye where the eyebrow is in humans. More commonly called a supercilium in birds.

Eye-stripe A coloured line running horizontally behind and in front of a bird's eye.

Foreneck The front section of the neck.

Frontal shield A hard plate extending upwards from the base of the maxilla onto the forehead and towards the crown.

Gape The 'hinge' of the bill.

Lesser Antillean endemic species A species that is found on multiple Lesser Antillean islands, but not found outside the region.

Lores The skin at the base of the bill.

Migrant A species that travels between countries and regions to breed and overwinter.

Migratory season The months during which North American breeding residents leave their breeding territories to travel south (often through the Lesser Antilles). Also the months during which these same birds return to North America from their overwintering grounds.

Morph A colour variation in plumage (typically pale or dark).

Maxilla The upper mandible of the bill.

Near-endemic species A species that can only be found on two islands in the Lesser Antilles.

Passerine Any bird that is specifically adapted for perching. Songbirds such as warblers, sparrows and tanagers are passerines.

Plumage A bird's feathering.

Primaries The main flight feathers – found towards the wing-tip.

Primary forest Mature, ancient forest often largely undisturbed by human influence or action.

Rufous Reddish-brown colouration.

Rump The area on the upperside of the body just above the tail.

Scrub Primarily dry habitat dominated by a mix of grasses, shrubs and short trees.

Spring migration The period spanning the months of March to April.

Subspecies Geographically separate populations of a species that can be genetically distinguished from each other.

Terminal band A colour band running across the tip of the tail.

Underparts The throat, breast, belly and vent.

Undertail The vent area.

Upperparts The nape, upper back, lower back, rump and uppertail.

Vent The area on the underside from which the bird expels waste.

Wing-bars Diagonal colour lines on the upperwing that are often (but not always) paler than the rest of the wing.

Xeric landscape Describes dry, often rocky, habitats.

BIBLIOGRAPHY

Allan, C. D., 2017. *Landscapes and Landforms of the Lesser Antilles*. Springer Nature, London.

Buckley, P. A., Massiah, E. B., Hutt, M. B., Buckley, F. G. & H. F., Hutt, 2009. *The Birds of Barbados: An Annotated Checklist*. British Ornithologists' Union, Oxford

Cornell Lab of Ornithology, 2021. https://birdsoftheworld.org/bow/home.

James, A., Durand, S. & Baptiste, B., 2005. *Dominica's Birds*. Forestry, Wildlife and Parks Division of Dominica.

Keith, A. R.,1997. *The Birds of St Lucia, West Indies: An Annotated Checklist*. British Ornithologists' Union, Oxford.

Kenefick, M., Restall, R. & Hayes, F., 2017. *Birds of Trinidad and Tobago*. Second edition. Bloomsbury Publishing, London.

Kirkconnell, A., Bradley, P. E. & Rey-Millet, Y. J., 2020. *Birds of Cuba: A Photographic Guide*. Bloomsbury Publishing, London.

Kirwan, G. M., Levesque, A., Oberle, M. & Sharpe, C. J., 2019. *Birds of the West Indies*. Lynx Edicions, Barcelona.

National Geographic Society, 2002. *Field Guide to the Birds of North America*. Fourth edition. National Geographic Society, Washington, D.C.

Raffaele, H. A. & Wiley, J. W., 2014. *Wildlife of the Caribbean*. Princeton Pocket Guides, New Jersey.

Raffaele, H. A., Wiley, J. W., Garrido, O. H., Keith, A. R. & Raffaele, J. I., 2020. *Birds of the West Indies*. Second edition. Princeton Field Guides, New Jersey.

Raffaele, H. A., Petrovic, C., Colón López, S. A., Yntema, L. D. & Salguero Faria, J. A., 2021. *Birds of Puerto Rico and the Virgin Islands*. Third edition. Princeton Field Guides, New Jersey.

Wege, D. C. & Anadon-Irizarry, V., 2008. *Important Bird Areas in the Caribbean: Key Sites for Conservation*. BirdLife International, Cambridge.

Wiley, J. W., 2021. *The Birds of St Vincent, the Grenadines and Grenada: An Annotated Checklist*. British Ornithologists' Club, Tring.

PHOTO CREDITS

Cover, front, top left to right: Lesser Antillean Euphonia (photo by Anthony Levesque), Red-legged Thrush (photo by Mark Greenfield), Purple-throated Carib (photo by Mark Greenfield); main photo: Martinique Oriole (photo by Béatrice Henricot); back, top to bottom: St Vincent Parrot (photo courtesy of St Vincent and the Grenadines Tourism Authority), White-breasted Thrasher (photo by Keith Clarkson), Bridled Quail-dove (photo by Béatrice Henricot), White-tailed Tropicbird (photo by Anthony Levesque).

All internal photographs taken by Yves-Jacques Rey-Millet with the exception of the following:

Key: t = top; b = bottom; l = left; r = right; c = centre; tl = top left; tr = top right; bl = bottom left; br = bottom right; cl = centre left; cr = centre right.

Faraaz Abdool: 24t, 24b, 26tr, 39t, 39b, 54b, 55t, 81b, 89b, 97b, 132b, 137t, 145t, 145bl, 145br, 147bl, 147br, 149t, 149b, 160t, 167b, 169b, 170t, 216t, 216b, 218t; Vaughan Ashby: 32t, 49cr, 50b, 52t, 69b, 92t, 92b, 94t, 96b, 100t, 102t, 108t, 127t, 127c, 128t, 129c, 133t, 151c, 152bl, 157tl, 159b, 165b, 170br, 173b, 175t, 175b, 177b, 217t, 217b; Mike Barth: 47b, 49b, 67t, 76t, 76br, 79tl, 81tr, 82t, 84b, 87t, 87c, 87b, 90t, 93b, 98b, 113b, 118t, 123bl, 123br, 174t, 176t, 199b, 203b; Alexandra Chenery: 21, 25b, 48bl, 74t, 75t, 75bl, 98c, 108c, 108b, 114b, 150br, 215bl; Ryan Chenery: 9–12, 25t, 28t, 28b, 29t, 36t, 36b, 41t, 41b, 43bl, 43br, 48t, 48br, 56tl, 56tr, 59tl, 64b, 74bl, 75br, 76bl, 79tr, 85cr, 85b, 90cr, 95tr, 95b, 114tr, 126tr, 126b, 130t, 147t, 150t, 150bl, 160b, 164t, 165t, 168br, 169t, 184t, 184bl, 184br, 185tl, 185tr, 198t, 211t, 212t, 212b, 213t, 215t, 215br, 218bl, 218br; Keith Clarkson: 34b, 59bl, 60t, 80br, 85t, 85cl, 132t, 137b, 146tr, 152tl, 153tl, 153bl, 153br, 154b, 155b, 161b, 162t, 162b, 163bl, 166t, 168t, 170bl, 179t, 183t, 205t, 210t, 211bl, 213br; John Dyson: 133bl, 143b, 211br; Andrea Easter-Pilcher: 131t, 131b; Vaughan Francis: 46b, 82b, 133br, 161t; Mark Greenfield: 57c, 172b; Skye Haas: 66t, 66b, 67bl, 67br, 68t, 68b, 69b, 70b, 128b; Jane Hartline: 61bl; Béatrice Henricot: 1, 18, 45t, 45b, 46t, 58t, 58b, 59tr, 114tl, 152tr, 152br, 153tr, 154t, 163t, 164b, 168bl, 171t, 172t, 181t, 181b, 182t, 182b, 191b, 198b, 204t, 205b, 210b, 213bl; Nigel Lallsingh: 52b; Alex Large: 3, 15, 26tl, 38t, 38bl, 38br, 43t, 49t, 56b, 59br, 64t, 65t, 65b, 116cl, 124t, 124b, 178br; Anthony Levesque: 17, 40b, 54t, 54c, 69t, 70t, 94b, 102b, 135t, 135bl, 135br, 138b, 171b, 176b, 179b, 191c; Julian Moore: 26b, 49cl, 83t, 83b, 95tl, 148tr; David Petts: 61t, 143t, 166b, 167tr, 183b, 214b; Steve Race: 78t, 78b, 177t, 178bl; Paul R. Reillo: 142t, 142b, 144l, 144r; St Vincent and the Grenadines Tourism Authority: 146tl, 146b, 191t, 219t; Larry Therrien: 148tl, 148b; Adams Toussaint: 214t; George Tuthill: 31t, 55b, 73t, 110tl, 173t, 174b, 219b; Alicia Williams: 61br, 115tl, 163br, 167tl; Steven Woon: 4, 178t.

INDEX

Anhinga 87
Ani, Smooth-billed 60
Avadavat, Red 176

Bananaquit 211
Bishop, Northern Red 173
Bittern, Least 72
Bobolink 180
Booby, Brown 86
Masked 87
Red-footed 86
Bullfinch, Barbados 212
Lesser Antillean 213
Bunting, Indigo 208

Carib, Green-throated 56
Purple-throated 57
Chachalaca, Rufous-vented 24
Coot, American 65
Cowbird, Shiny 185
Cuckoo, Mangrove 61
Yellow-billed 60

Dove, Common Ground 49
Eared 49
Eurasian Collared 41
Grenada 46
Rock 41
White-winged 47
Zenaida 48
Dowitcher, Long-billed 107
Short-billed 106
Duck, Masked 29
Muscovy 31
Ring-necked 31
Ruddy 30
Dunlin 99

Egret, Cattle 76
Great 79
Little 83
Snowy 82

Elaenia, Caribbean 148
Yellow-bellied 147
Euphonia, Lesser Antillean 179

Falcon, Peregrine 141
Finch, St Lucia Black 214
Flycatcher, Fork-tailed 151
Grenada 153
Lesser Antillean 152
Frigatebird, Magnificent 85

Gallinule, Purple 63
Godwit, Hudsonian 94
Marbled 94
Grackle, Carib 184
Grassquit, Black-faced 215
Blue-black 216
Grebe, Pied-billed 38
Grosbeak, Rose-breasted 207
Guineafowl, Helmeted 24
Gull, Black-headed 113
Herring 116
Laughing 114
Lesser Black-backed 116
Ring-billed 115

Hawk, Broad-winged 133
Common Black 132
Red-tailed 134
Hermit, Rufous-breasted 55
Heron, Great Blue 77
Green 75
Grey 78
Little Blue 81
Purple 78
Tricolored 80
Hummingbird, Antillean Crested 59
Blue-headed 58

Ibis, Glossy 71

Jaeger, Arctic 128
Long-tailed 127
Pomarine 127
Junglefowl, Red 25

Kestrel, American 139
Killdeer 91
Kingbird, Grey 150
Tropical 149
Kingfisher, Belted 136
Ringed 135
Kite, Hook-billed 131
Knot, Red 96

Lapwing, Southern 92

Martin, Caribbean 160
Merlin 140
Mockingbird, Tropical 167
Moorhen (Gallinule), Common 64
Munia, Chestnut 177
Scaly-breasted 177
White-headed 178

Nighthawk, Antillean 51
Common 50
Night-heron, Black-crowned 73
Yellow-crowned 74
Nightjar, Rufous (St Lucia) 52
White-tailed 52
Noddy, Brown 112

Oriole, Martinique 181
Montserrat 182
St Lucia 183
Osprey 130
Ovenbird 185
Owl, Ashy-faced 129
Oystercatcher, American 88

Parakeet, Rose-ringed 147
Parrot, Imperial 144
Orange-winged 145
Red-necked 142
St Lucia 143
St Vincent 146
Parula, Northern 195
Pelican, Brown 84
Petrel, Black-capped 67
Pewee, Lesser Antillean 154
Phalarope, Wilson's 108
Pigeon, Scaly-naped 43
White-crowned 42
Pintail, Northern 37
White-cheeked 36
Plover, American Golden 89
Collared 92
Grey 89
Semipalmated 90
Snowy 91
Wilson's 90

Quail-dove, Bridled 45
Ruddy 44

Rail, Clapper 62
Redstart, American 193
Ruff 96

Saltator, Lesser Antillean 210
Sanderling 98
Sandpiper, Baird's 100
Buff-breasted 102
Least 100
Pectoral 103
Semipalmated 104
Solitary 109
Spotted 108
Stilt 97
Upland 93
Western 105
White-rumped 101
Scaup, Greater 32
Lesser 32
Seedeater, Yellow-bellied 217
Seed-finch, Chestnut-bellied 217
Shearwater, Audubon's 70
Cory's 69
Great 68
Manx 70
Sooty 69
Shoveler, Northern 33
Skua, Arctic 128
Long-tailed 127
Pomarine 127
South Polar 128
Snipe, Wilson's 107
Solitaire, Rufous-throated 168
Sora 62
Sparrow, House 178
Stilt, Black-necked 88
Storm-petrel, Leach's 67
Wilson's 66
Swallow, Bank 159
Barn 158
Cave 157
Cliff 157
Swift, Black 53
Grey-rumped 54
Lesser Antillean 54
Short-tailed 55
White-collared 53

Tanager, Lesser Antillean 219
Scarlet 209
Teal, Blue-winged 34
Tern, Arctic 124
Black 123
Bridled 118
Caspian 121
Common 123
Gull-billed 120
Least 119
Roseate 122
Royal 126
Sandwich 125
Sooty 117
Thrasher, Pearly-eyed 164
Scaly-breasted 163
White-breasted 162
Thrush, Cocoa 169
Forest 171
Red-legged 172
Spectacled 170
Toucan, Channel-billed 137
Trembler, Brown 165
Grey 166
Tropicbird, Red-billed 39
White-tailed 40
Turnstone, Ruddy 95

Vireo, Black-whiskered 155
Red-eyed 155
Yellow-throated 156

Warbler, Barbuda 204
Blackpoll 200
Black-throated Blue 201
Black-throated Green 206
Black-and-White 187
Canada 204
Cape May 194
Chestnut-sided 199
Hooded 192
Kentucky 189
Magnolia 196
Palm 202
Plumbeous 191
Prairie 203
Prothonotary 188
St Lucia 205
Whistling 191
Yellow 197, 198
Waterthrush, Louisiana 186
Northern 186
Waxbill, Black-rumped 175
Common 176
Orange-cheeked 175
Weaver, Village 174
Whimbrel 93
Whistling Duck, Black-bellied 26
Fulvous 28
West Indian 27
Wigeon, American 35
Willet 110
Woodpecker, Guadeloupe 138
Wren, House 161

Yellow-finch, Grassland 218
Yellowlegs, Greater 111
Lesser 111
Yellowthroat, Common 190